T0254432

NEXUS NETWORK JOURNAL

MATHEMATICAL PRINCIPLES IN ARCHITECTURAL DESIGN

VOLUME 8, NUMBER 1

Spring 2006

KIM WILLIAMS BOOKS

Nexus Network Journal
Vol. 8
No. 1
Pages 1–139
ISSN 1590-5896

Contents

Editorial
1   Letter from the Editor
       K. WILLIAMS

Research Articles
5   The Acropolis of Alatri: Architecture and Astronomy
       G. MAGLI

17   The Stylistic Characteristics of the Shampay House of 1919: A Formal Analysis
       J. H. PARK

33   Timely Timelessness: Traditional Proportions and Modern Practice in Kahn's Kimbell Museum
       S. FLEMING, M. REYNOLDS

53   Origins of an Obsession
       D. J. MARSHALL II

The Geometer's Angle
67   The Golden Section
       R. FLETCHER

Didactics
93   Mathematical Mode of Thought in Architectural Design Education: A Case Study
       I. M. VERNER, S. MAOR

107   An Introduction to Algorithms and Numerical Methods using Common Software
       J. BRANGÉ

112   Natural Architecture and Constructed Forms: Structure and Surfaces from Idea to Drawing
       M. ROSSI

123   Sound-Sights - An Interdisciplinary Project
       C. LEOPOLD

Book reviews
135   Misteri e scoperte dell'archeoastronomia
       M. INCERTI

138   Coincidence, Chaos and All That Math Jazz
       K. WILLIAMS

LETTER FROM THE EDITOR

*F*rom ancient to modern, architects have looked for fundamental underlying principles of geometry and proportion on which to found their designs. Such principles not only provide an order for the formal elements, they ground the architecture in timeless values and provide a source of cultural meaning. This issue of the *Nexus Network Journal* illustrates the use of such principles in two ancient cultures, the Bronze Age and the Roman, as well as in twentieth-century North America.

In "The Acropolis of Alatri: Architecture and Astronomy" Giulio Magli examines the construction techniques of the polygonal walls of this ancient city, and its alignment with the position of the stars at the time of construction, shedding light not only on the history of construction, but on the history of science as well. Using techniques of archaeoastronomy and pointing out that stars in constellations no longer visible today were visible thousands of years ago, Magli is able to sustain relationships between the site and the ancient heavens, and make evident the ancient builders' sense of connection between the terrestrial and celestial in their architecture and urban planning.

Obsessed with geometry, D.J.P. Marshall undertook a geometrical experiment to reconstruct the design of the Forum of Augustus with the classical instruments of straightedge and compass. Basing his procedure on the simplest manipulations of the square, Marshall was able to account convincingly for the positions of the major elements of the Forum. While we know that such reconstructions are speculative, and that geometrical and proportional relationships can exist in a work of built architecture as a consequence of design decisions rather than as initial input, Marshall's step-by-step geometrical procedure is a model of how geometrical design develops and just how effective it is as an ordering device. Simple geometrical forms such as the square were also powerful symbols for natural laws.

We may tend to think that laws of proportion and geometry appertain more to the past than the present, but the two studies of twentieth-century architects Louis Kahn and Rudolph Schindler both show how timeless these sciences are for modern architecture.

Jin-Ho Park's formal analysis of the Shampay House of 1919 illustrates the care with which Schindler considered spatial arrangement, symmetry and proportion. A thorough understanding of mathematical concepts such as axial arrangements allowed Schindler to create a design that was well-ordered yet flexible. Along with Lionel March, Park has discussed Schindler's proportional concept of the "row" in a previous issue of this journal (see *NNJ* vol. 5 no. 2, Autumn 2003). The present paper tells us more about Schindler's rich and varied "toolkit" of design principles.

Architecture historian Steven Fleming and geometer Mark Reynolds teamed up to analyze the geometry and proportions of the Kimbell Art Museum by one of the United States's most enigmatic architects, Louis Kahn. Kahn tantalizes and frustrates by hiding or denying his design principles in the written documents that have come down to us, but the fruitful collaboration of Fleming and Reynolds is very revealing. Kahn cannily reconciles timeless architectural design principles with the requirements and limits of his clients. Nexus can take some credit for this, because the two scholars met at a Nexus conference, and their collaboration represents an ideal of the Nexus community, where practitioners of different disciplines find a window of opportunity to exchange ideas and points of view that expand their respective horizons.

Geometer Rachel Fletcher gives an important lesson in the most classic and most discussed of all geometrical design tools, the Golden Section. Used as a guiding principle in architecture both

1590-5896/06/010001-2 DOI 10.1007/s00004-006-0010-1

ancient (as in the Forum of Augustus) and modern (as in Kahn's Kimbell Museum), the Golden Section is beautiful both mathematically and aesthetically.

When in the first book of his *Ten Books* Vitruvius describes the preparation of an architect, he says:

> Let him be educated, skilful with the pencil, instructed in geometry, know much history, have followed the philosophers with attention, understand music, have some knowledge of medicine, know the opinions of the jurists, and be acquainted with astronomy and the theory of the heavens.

The domains of knowledge of today's architect are perhaps different but no less varied. The task of the educator is to teach the student even in the face of the difficult question, "But why should I learn mathematics?"

One of the answers to this question is that mathematics is present in the natural world. Michela Rossi describes that in the course that she teaches on "Natural Architecture and Constructed Forms." The comprehension of geometric schemes in regular organic objects form the basis of teaching drawing and scientific representation, such as formal architectural synthesis. This exercise may also offer a valid starting point to help students approach mathematics, and to help them imagine and plan the increasingly complex surfaces of contemporary architecture.

Skill in draftsmanship has been replaced by skill in CAAD. The powerful tool given to architects by information science has expanded the possibilities but requires other knowledge as well. In "An Introduction to Algorithms and Numerical Methods Using Common Software," Jean Brangé describes an approach using software applications that students are already familiar with, such as Photoshop, VRML and C4D, to manipulate geometric and algebraic formulas, recursion, random functions, statistics, splines, the fourth dimension and other complex mathematical concepts

Igor Verner and Sarah Maor have given considerable thought to the various theories about teaching mathematics to architecture students, including the "Realistic Mathematics Education" approach and the "Mathematics as a Service Subject" approach. In "Mathematical Mode of Thought in Architecture Design Education: A case study," Verner and Maor present their own experience with designing and implementing mathematics courses for architecture students.

Cornelie Leopold involved her students with the concepts of geometry, architecture and music by engaging their senses. In an project entitled "Sound–Sights" involving students of architecture and musical composition, interdisciplinary groups of students were asked to give architectural form to musical ideas, and musical form to architectural compositions. The result was an exhibit and performance at a concert hall in Kaiserslautern.

This issue concludes with two book reviews. Manuela Incerti reviews *Misteri e scoperte dell'archeoastronomia. Il potere delle stelle dalla preistoria all'isola di Pasqua* by Giulio Magli. Prof. Magli's book, in Italian, provides the background for his article in this issue. Kim Williams reviews *Coincidence, Chaos and All That Math Jazz* by Edward B. Burger and Michael Starbird. There are many examples here of the power and omnipresence of mathematics in our everyday world, including art and architecture.

Kim Williams

Research

Giulio Magli

Dipartimento di Matematica
Politecnico di Milano
P.le Leonardo da Vinci 32
20133 Milano, Italy
magli@mate.polimi.it

The Acropolis of Alatri: Architecture and Astronomy

The astronomical alignments of the Acropolis of Alatri, Italy, are investigated. The results strongly support a dating of the magnificent polygonal walls of the site to a pre-Roman period.

1 Introduction: polygonal walls in Italy

The so-called *cyclopean* or *polygonal* walls are huge walls of megalithic blocks cut in polygonal shapes and fitted together without the use of mortar. Many walls of this kind were constructed during the Bronze Age in the Mediterranean area (probably the most famous are the walls of Mycenae; see e.g. [Heizer 1990]). Slightly less known but equally impressive and magnificent are the walls visible in many Italian towns, spread over a wide area which spans from Umbria to Campania. Besides Alatri, which will be the subject of the present paper, those of Segni, Ferentino, Norba, Arpino and Circei are worth mentioning.

Polygonal walls are traditionally classified into three types or *manners*. The first manner is constructed with blocks of medium dimensions with coarse and inaccurate joints; the second manner shows good stonework with few wedge fillings between main blocks; the third manner is simply perfect, with joints so accurately prepared that it is impossible to insert even a sheet of paper or the blade of a pocket knife between two adjacent stones. The small town of Alatri, in the Frosinone provincial district about an hour's drive from the center of Rome, is the place where this third manner of polygonal walls is still visible nearly intact, as it was constructed at least 2400 years ago. The masonry of Alatri is nearly identical to that of the famous polygonal wall of the Greek Delphi sanctuary, a wall dated no later than circa 520 BC. The two are so similar that one actually gets the impression that they were built by the same engineer (fig. 1).

Fig. 1a Polygonal walls of the Alatri Acropolis. Porta Minore is visible on the right

Fig. 1b The polygonal wall at Delphi

2 The Alatri Acropolis

The city of Alatri was built around a small hill, and the town is surrounded by megalithic walls of which many remains are still visible today. The Acropolis in turn is a gigantic construction built on the hill and covering the top of it. In some sense the hill was adapted and sculpted in such a way as to obtain a sort of "geometric castle" dominating the center of the town (fig. 1a). The Acropolis is so impressive that the famous German historian Gregorovius (1821-1891) reported "an impression greater than that made by the Coliseum".

The perimeter of the building is defined by huge walls which give to it a polygonal shape with six sides ABCDEF (fig. 2). The polygonal shape of the Acropolis actually looks like a giant replica of one of the stone blocks of which the Acropolis itself is constructed. The access to the Acropolis was possible trough two doors, today called Porta Minore (fig. 3) and Porta Maggiore (fig. 4). Porta Minore is a small trilithon doorway, and on its lintel a symbol composed by three phalli disposed as to form the upper part of a crux can be discerned (fig. 5).

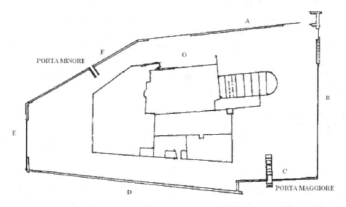

Fig. 2 .Plan of the Acropolis in Alatri

Fig. 3. Porta Minore

Fig. 4. Porta Maggiore. The staircase was added in the eighteenth century

Fig. 5. The lintel of Porta Minore, with the half-crux phallic symbol

Porta Maggiore is one of the most magnificent megalithic structures in Europe, and it is composed of a tunnel of huge stone blocks with a corbelled ceiling and a monumental trilithon access. After the door on the C wall, a short bent wall leads to side D, which contains three huge "niches". These "niches" look like basements for statues (fig. 6), but no kind of artistic find has been found here or elsewhere in the Acropolis. Thus, the unique "message" left today by the builders of this monument is the half-crux-shaped phallic symbol on the small door (a statue of a lion, badly damaged, lies on the ramp on the north side, but the dating of this sculpture is unknown).

Fig. 6. Two of the three niches on wall D

On the top of the Acropolis an additional megalithic structure, lying on a natural rocky platform, existed. Archaeologists call this structure *ierone*, thinking that it was the basement of a temple. However, no proof of this statement has ever been found. The structure was constructed with enormous stone blocks perfectly cut and joined, without mortar, in such a way that it is impossible, as mentioned, to insert even a sheet of paper between two blocks (fig. 7). On this structure the Alatri cathedral was later built, so that today only the lower courses of stones remain visible under the church.

Fig. 7. The huge polygonal basement on the rocky terrace of the Acropolis,
which today lies under the northern side of the Cathedral

3 Solar alignments and geometrical symmetries in the Alatri construction plan

The first to propose the idea that the city of Alatri and its Acropolis were planned on the basis of geometrical and astronomical alignments was a local historian, Don Giuseppe Capone, in 1982. His work is poorly known and of very difficult to access, and therefore I will review here the most important discoveries of this scholar ([Capone 1982, 12-14]; see also [Aveni 1985] where a brief account in English is given).

Capone studied the geometry of the city plan and discovered that there exists a sort of "privileged point" (indicated by O in figs. 2 and 8) which lies just behind the northern wall of the megalithic structure at the center of the Acropolis (and thus today lies near the northern wall of the Church, see fig. 7). With respect to point O, Capone identified many geometrical symmetries and astronomical alignments (fig. 8):

- The northeast (H) corner at of the Acropolis defines a direction OA which identifies the rising sun at the summer solstice;

- The eastern and western sides of the Acropolis are parallel and oriented north-south;

- The city has six main doors (indicated by P1-P6) and three small doors or *portelle* (p1-p3). The north-west sector has two main doors and two small ones, the north-east sector two main doors. "Symmetrically dividing by two" with respect to O, the south-west sector has one main door and the south-east sector one main door and one small door;

- The lines connecting the couples of doors p1-p3, p2-P4, P3-P5 intersect each other in O;

- The p1-p3 line is perpendicular to BA ;

- All the main doors except P2 are equidistant from O. The distance equals three times the value of the segment OH (which is about 92 m. long). This value seems to have governed the whole planning of the city.

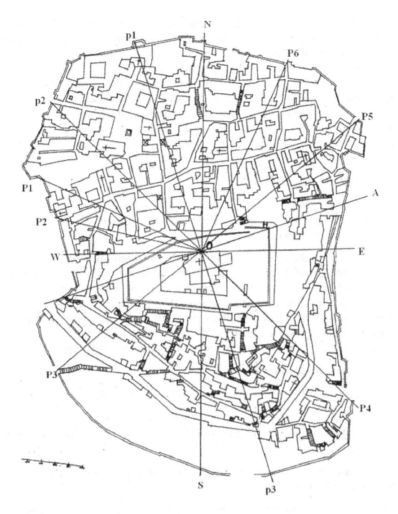

Fig. 8 Plan of Alatri, with the alignments discovered by G. Capone (adapted from Capone 1982)

Capone was also puzzled about the shape of the Acropolis, which is *not* governed by the morphology of the hill. As a matter of fact, it is rather the contrary, because the hill was adapted, sculpted, in such a way so as to obtain the desired shape. The idea that Capone courageously suggested was that the plan of the Acropolis could have been conceived as an image of the Gemini constellation.

It should be noted that Capone's idea was greatly original. In fact, today we do know of many examples where people of the past tried to connect earth and heavens, and even to replicate the sky on the ground, by means of astronomically-related buildings. Some such examples are controversial and not all scholars accept them – for instance, the theory that interprets the disposition of the three main pyramids in Giza as a representation of the three stars of Orion's belt [Bauval1989] – but others are certain, such as the use of *hyerophanies*, "sacred machines," which were activated by specific celestial events; among them, the famous *Castillo* of Chichen Itza, in the

Yucatan, a Toltec-Maya pyramid constructed in such a way that a light-and-shadow serpent descends its staircase at the equinox [Aveni 2001]. Another example is that of the megalithic temples of Malta. These huge buildings, constructed between 3500 and 2500 BC, were planned according to a complex cosmographic concept, which included the "shape" of the so-called "mother goddess" (a feminine "fat" deity) in the internal layout of the temples, the orientation of the main axis to the rising of the Southern Cross-Centaurus asterism [Hoskin 2001] and probably also the orientation of the left "altar" of the temple to the winter solstice sunrise [Albrecht 2001]. Recently, I proposed a similar "cosmographic principle" for the Inca capital, Cusco; according to this proposal, the city could have been laid out as a *replica* of a dark cloud constellation having the form of puma [Magli 2005].

4 Stellar alignments of the Acropolis

I present here the results of an investigation of the possible astronomical content of the planning of the Acropolis. The data for the alignments of the walls were collected with a precision which can be reasonably assumed to be around 0.5°.

The idea that the plan of the Alatri Acropolis could have been governed by further stellar alignments came to me in light of the fact that the eastern and western sides are oriented cardinally (they deviate by 0.5° and 0.8° west of north respectively), and that the first available "angle" for a sight line is oriented to the summer solstice sunrise. Why, then, did they not construct the southern side with squared angles as well? Further, why did they not use squared angles, or a single bent line, for the northern side? As we have seen, the Acropolis was not constructed with a form adapted to the shape of the hill, and therefore geo-morphological reasons have to be excluded.

The orientations of the sides are the following:

- The two southern walls are oriented about 4.7° west of south (C side) and about 3.4° east of south (D side);

- The two northern walls are oriented about 28° west of north (E side) and about 12.6° west of north (F side).

To search for an astronomical answer to this puzzle I will take into consideration the scant traces which the builders left to us.

There are only two exceptions to the complete, frightening silence left by them, and both have something to do with a symbolism in which the number three played some role: three phalli forming the half-crux on the Porta Minore, and three huge niches on wall D.

Of course, there is a very important crux in the sky, the group of stars which from the sixteenth century has been known as the Southern Cross. Today, due to precession, it is invisible in Italy, but these stars are currently in the lower part of their precessional cycle and they were actually visible in the Mediterranean area in ancient times. The importance of the asterism composed by the bright stars of Centaurus and the stars of Crux is very well documented not only in the previously-cited megalithic temples of Malta (3400-2000 BC) but also in the megalithic sanctuaries of Minorca of around 1500-1000 BC [Hoskin 2001]. In addition, it has also recently been found in the orientation of the Sardinian towers called Nuraghes (around 1200-800 BC) [Zedda and Belmonte 2004].

If we take a look at the southern sky in Alatri at a reference date of, say, 400 BC we discover that the Southern Cross was only partly visible. In fact, the lower star of the Southern Cross,

Acrux, culminated below the southern horizon and therefore was not visible: the "cross" actually appeared as a "half-cross" composed of *three* stars (actually four if one considers the star *epsilon-crucis*). Moreover, of the two brightest stars of Centaurus, Hadar and Rigel, only Hadar was still visible at that time.

The azimuth at the rising and setting of all such stars cannot of course match the directions very close to due south defined by the aforementioned azimuths of the southern sides of the Acropolis (4.7° west of south and 3.4° east of south). However if we consider *at the same time* both such alignments and suppose that the side containing the three niches was oriented to a position of Crux, then it can be seen that, when the southernmost star of Crux had an azimuth of about 4.7° west of south, the bright star Hadar had an azimuth of 3.4° east of south. Therefore, I suggest that the southern sides of the Acropolis could have been oriented to the Southern Cross–Centaurus asterism in a position when the stars were in the configuration depicted in fig. 9. Of course due to the slowness of the precessional drift, the date has to be considered only as indicative. Since the star Acrux became invisible at the Alatri latitude around 700 BC, this latter date has to be considered as a *post-quem* term for the present theory, which therefore puts the planning of the Acropolis between the seventh and the fourth centuries BC. However, since this kind of interest in the southern sky is not documented by the Romans, this astronomical orientation of the southern sides of the Acropolis, if confirmed, would connect the builders of Alatri to the widespread tradition of astronomical observations of the Southern Cross-Centaurus group of the Mediterranean area in the Bronze Age, in the meanwhile suggesting a pre-Roman dating of the Acropolis.

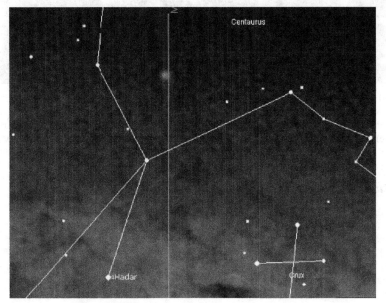

Fig. 9. The sky close to due south in Alatri in 400 BC

We now turn to the northern sky. Looking in the region around 30° west of north, one immediately sees that Capella, a bright star whose importance is well documented in many cultures throughout the world, was setting at about 32° west of north as viewed from Alatri in 400 BC. The bright stars Vega and Deneb also had setting azimuths in the same region (around 34° for both). However, Capella seems preferable, both because it is the star nearest to the desired direction and because of the following observation.

As I have mentioned, the Italian scholar Capone noticed a similarity between the plan of Alatri and the constellation Gemini, and proposed that this could have inspired the planning of the site. Actually if we look at the northern sky in Alatri at our reference date of 400 BC, we discover that while Capella was setting, *Gemini was also setting* and this constellation was standing as a "polygon" with a shape that is actually very similar to that of the Acropolis, with the side "looking" at Capella in correspondence with the north-west "bent" side of the Acropolis (fig. 10).

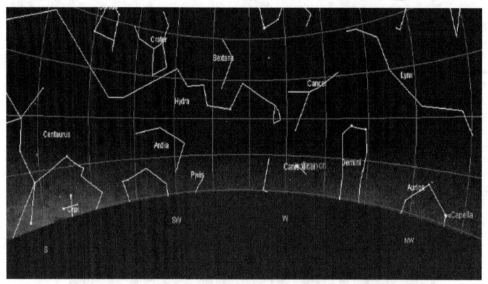

Fig. 10 The western part of the sky as viewed from Alatri in 400 b.C. From left to right: on the two "sides" of the south pole, the Crux-Centaurus group with the bright star Hadar east of south and the Crux west of south. The lower star of Crux was invisible (not only in this picture) due to precession. At the northwest horizon, we see Gemini in a configuration very similar to the Acropolis plan. The "bent" side of Gemini "looks" at the bright star Capella setting at about 30° west of north

5 On the dating of Italian polygonal walls

The builders of the polygonal walls in Italy have *not* been identified with certainty.

A long-standing debate divides those who think that most of the walls were built by the Romans or by Latin people under the strict and direct influence of the Romans, and who therefore date the most ancient of these huge constructions no earlier than the fourth century BC, and those who think that Romans simply had nothing to do with such buildings and therefore that the walls can be several hundred years older than this.

The question about the builders of the Italian polygonal walls was debated at the beginning of the twentieth century, when most scholars adhered to a long-standing tradition which attributed the construction of the walls to a people called *Pelasgi*, who allegedly came to Italy at the end of the Bronze Age (1200-800 BC), bringing with them Hittite-Mycenaean technology. However, proof of the existence of this people is scant and, as a matter of fact, the diffusion of techniques need not necessarily be identified with the diffusion of a population bringing it, since a technique can be imported via cultural and economic contacts or can be independently invented (it suffices to think of the magnificent polygonal walls built by the Incas two thousand years later). Thus, the walls might be attributed as well to the populations inhabiting the areas before the Roman expansion, people who, like their contemporaries the Etruscans who lived in adjacent areas, had active economic and cultural contacts with the civilizations of the Mediterranean. In 1957 however, this explanation was refuted by the authoritative scholar Giuseppe Lugli [1957] who re-formulated, this time with an apparently unassailable argument, the idea that the polygonal walls were nothing but one of the types of Roman walls, a type called *by him* – I repeat, *by him* – *opus siliceum*, to be *added* to the already well known types of Roman walls described by Vitruvius such as, for instance, *opus quadratum*, walls of stone blocks cut as parallelepipeds and set in place on horizontal layers without mortar, and *opus caementicium*, in which the core of the walls is made of pieces of stones mixed with mortar and sand, while the external faces are of small stones or bricks.

Recently, many scholars of the builders of the polygonal walls are again becoming more doubtful[1] and the important problem of the dating of the walls deserves in-depth treatment in a separate publication. However, the dating of the Alatri walls is of fundamental importance for the present paper, since the results support a pre-Roman construction. Therefore, a discussion of the topic is necessary.

Dating of stone building *per se* is impossible.[2] While it is common to find on Roman bricks or even on Roman square blocks the presence of marks and timbres identifying the "producers" of the material and in many cases also the year of its extraction from the cave, no traces of this kind have ever been found on polygonal blocks anywhere in Italy. Actually, Lugli's argument of a Roman dating was founded only on a very debatable interpretation of a statement by Vitruvius (*De architectura* 1.5.II), who *cites* the existence of a sort of construction called *silice* without actually describing it, and on a even more debatable interpretation of a short inscription (dated around 140 BC) which runs on the walls of the Ferentino acropolis. In this inscription, the *censors* (magistrates) who certainly built the upper part of the construction in square blocks, where the inscription is actually located, give the impression of crediting themselves also for the construction of the lower part of the same building, in polygonal blocks, calling it, again, *silice*. The text is, however, quite ambiguous [Solin 1980-82, 102 and references therein].

Against Lugli's argument many points can be raised, of which the main ones are:

- Romans were, of course, skilled in working with huge objects, such as the granite columns of the Pantheon in Rome, weighing more than two hundreds tons and coming from southern Egypt. They were skilled in building with huge blocks as well, as one can easily see by visiting, for instance, the internal rooms of Castel Sant'Angelo in Rome, which were originally built as the funerary chambers of the mausoleum of the emperor Hadrian. However, *all* the stone buildings *in Rome* were constructed with square blocks (*opus quadratum*) from the very beginning of the Roman civilization, as one can see, for instance, from the remains of the archaic defensive walls of the town, on the Palatine Hill, dated around 530 BC, or from the internal walls of the prison called *Carcer Tullianum*, which were built between the sixth and the fifth century BC, not to mention the first complete defensive walls of Rome, called *serviane*, dated around 390 BC. The fact that polygonal walls were never built in Rome is sometimes explained by observing that the quality of stone most easily available *near* the town, called *tufa*, is not easily cut in polygonal blocks (see e.g. [Guadagno 1988]). If this argument can perhaps explain the most ancient constructions in Rome, it is, at least in my opinion, really flawed insofar as the civilization destined to become the owner of the whole western world is concerned.

— The megalithic style of the polygonal walls (with blocks that can be as heavy as thirty tons) requires techniques which are completely different from those of *opus quadratum*, in which even the hugest blocks weigh at most a few tons. Romans made large use of tackles and pulleys, but stones like those visible in Alatri, Segni and many other places can be efficiently raised only with the help of earth ramps (perhaps one can raise *one* stone of this kind with the help of an appropriate system of pulleys, but building kilometres of walls with such devices appears impossible). In addition, before putting third-manner polygonal blocks *in situ* the hard work of fine-cutting is required so that the angles of the stones already positioned fit perfectly, while square blocks can be used exactly as they arrive from the cave. Once again, it is frankly difficult to believe (or at least, it is difficult for the present author) that such a spectacular and sophisticated technique was exclusive to the provinces.

— Techniques based on huge blocks were used by the Romans for the retaining walls of earthworks in suburban "villas" (farms) of the Republican age (second-first century BC). However, such walls cannot be classified into one of the three manners and, in fact, in order to describe them Lugli was obliged to introduce an *ad hoc* a "fourth manner" described as "very near to the square blocks technique". In the walls constructed with this "fourth manner" – actually a sort of "missing link" of Lugli's argument – there is a clear tendency to use horizontal setting layers, although the disposition is inaccurate and some blocks are trapezoidal and exhibit acute angles. This style is actually the *only* manner used in most, if not in all, the foundations of villas and, as a matter of fact, rather than considering it an "evolved" polygonal blocks technique it could be safely (and very reasonably) interpreted as a naïve way of building with square blocks, which was adopted because the construction did not require aesthetic beauty (for instance, the walls were used for agricultural terraces). The construction technique of the walls of the Acropolis in Ferentino could perhaps be included in this fourth manner as well, thereby validating the credits of the two censors.

— If the Romans used such complicated megalithic techniques, it is strange that they did not leave any document describing or at least depicting it, even in some disguise. There is no Roman historian citing them, not even Livius, who described the foundation ("deduction") of many colonies where, according to Lugli's argument, polygonal walls were constructed immediately after the settling of the colonists. We do have many stelae and reliefs illustrating buildings or even construction techniques of buildings, but in *all* such documents the blocks shown are regular, squared blocks.

— Another point which, at least in this author's view, shows how far the "Roman argument" is from the truth is that the megalithic builders did not use the arch. Their "arches" are trilithon doorways, or the so-called "false arches" (a very bad terminology), composed of corbelled blocks forming an upside down V. In contrast, Roman architects used the arch from the very beginning of the Roman civilization. As a consequence, one can see Roman restoration and integration of polygonal walls in which previously existing "false arches" were substituted by "true arches".

As we have seen, those who refute Lugli's argument (including myself) usually think that the construction of most of the walls pre-dates the Romans and can be attributed to the populations inhabiting the area between the eighth century BC and the Roman expansion in the fourth century. However, only a few sites have actually been securely dated so far (with the use of organic material or pottery associated with the walls) and some of them turned out to belong to such a period.[3] Further archaeological work in this direction is certainly advisable, especially in light of the astronomical facts considered above, and the old idea of a Bronze Age retro-dating of the walls cannot be ruled out completely at present. As a matter of fact, it has already happened in the history of archaeology that the dating of a stone building had to be shifted back in time of as much as 1000 years (I am, of course, referring to the Sarsen phase of Stonehenge, today dated around 2100 BC; see e.g. [Pitts 2001]).

Acknowledgments

I am indebted to Dr. Antonio Agostini, director of the Library of the Alatri Administration Municipality, for his kind help in providing the material on which this paper is based, and with Prof. Gian Luca Gregori of the University of Rome for his kind help in providing references on the Ferentino inscription. Warm thanks goes also to dr. Roberto Giambò for many discussions.

Notes

1. See e.g. [Marta 1990], who states "It is a characteristic of Italic civilization before Roman period: it goes from the end of the VI century to the I century BC."
2. Recently, a thermoluminescence method which allows the dating of the time at which a stone belonging to a wall has been cut has been developed and applied as a test-case to the polygonal wall at Delphi and to other stone buildings in Greece; see [Liritzis 1994], [Liritzis et al. 1997] and [Theocaris et al. 1994]. Of course, it would be worthwhile to apply such a method to the Italian cyclopean walls as well.
3. One the few sites which have been securely dated is the series of megalithic terraces located at the site of *Norba-Monte Carbolino*, which were constructed in the sixth century BC.

References

ALBRECHT, K. 2001. *Maltas tempel: zwischen religion und astronomie*. Potsdam: Naether-Verlag.

AVENI, A.F. 1985. Possible astronomical reference in the urbanistic design of ancient Alatri. *Archaeoastronomy* **8**, 1: 12-15.

———. 2001. *Skywatchers: A Revised and Updated Version of Skywatchers of Ancient Mexico*. Austin: University of Texas Press.

BAUVAL, R. 1989. A master plan for the three pyramids of Giza based on the three stars of the belt of Orion. *Disc. Egypt* **13**: 7-18.

GUADAGNO, G. 1988. Centosessanta anni di ricerche e studi sugli insediamenti megalitici: un tentativo di sintesi. In 1° Seminario Nazionale di Studi sulle *Mura Poligonali*, F. Cairoli Giuliani, ed. Alatri.

CAPONE, G. 1982. *La progenie hetea*. Alatri.

HEIZER, R.F. 1990. *The age of the giants*. Venice: Marsilio-Erizzo.

HOSKIN, M. 2001. *Tombs, temples and their orientations*. Bognor Regis: Ocarina Books.

LIRITZIS, I. 1994. *A new dating method by thermoluminescence of carved megalithic stone building*. Paris: Comptes Rendus (Academie des Sciences).

LIRITZIS, I., P. GUIBERT., F. FOTI and M. SCHVOERER. 1997. The Temple of Apollo (Delphi) strengthens new thermoluminescence dating method. *Geoarchaeology International* **12**, 5: 479-496.

LUGLI, G. 1957. *La tecnica edilizia romana*. Rome: Bardi.

MAGLI, G. 2005. *Mathematics, astronomy, and sacred landscape in the Inka heartland*. Nexus Network *Journal* **7**, 2: 22-32.

MARTA, R. 1990. *Architettura Romana – Roman Architecture*. Rome: Kappa.

PITTS, M. 2001. *Hengeworld*. London: Arrow.

SOLIN, H. 1980-82. Le iscrizioni antiche di Fermentino. *Rend. Pont. Acc. Rom. Arch.* **53-54.**

THEOCARIS, P., I. LIRITZIS and R.B. GALLOWAY. 1994. *Dating of two Hellenic pyramids by a novel application of thermoluminescence. Journal of Archaeological Science* **24**: 399-405.

ZEDDA, M. and J. BELMONTE. 2004. On the orientation of sardinian nuraghes: some clues to their interpretation. *J. Hist. Astr.* **35**: 85-102.

About the author

Giulio Magli is a full professor of Mathematical Physics in the Faculty of Civil, Environmental And Land Planning Engineering of the Politecnico of Milan, where he teaches courses on Differential Equations and Rational Mechanics. He earned a Ph.D. in Mathematics at the University of Milan in 1992 and his research activity developed mainly in the field of General Relativity Theory, with special attention to problems of relevance in Astrophysics, such as stellar collapse. His research interests include History of Astronomy and Archaeoastronomy, with special emphasis on the relationship between architecture, landscape and the astronomical lore of ancient cultures. On this subject he recently authored the book *Misteri e scoperte dell'archeoastronomia* (*Mysteries and Discoveries of Archaeoastronomy*), published by Newton & Compton.

Jin-Ho Park

Department of Architecture
Inha University
253 Yonghyun-dong, Nam-gu,
Incheon
402-751 Korea
jinhopark@inha.ac.kr

The Stylistic Characteristics of the Shampay House of 1919: A Formal Analysis

This paper analyzes the stylistic characteristics of the Shampay House with a series of formal methodologies. It focuses on three parts: spatial arrangement, symmetry and proportion. For the thorough analysis, archival drawings are enhanced through reconstructing new drawings and through the building of a quarter-inch scale model.

1 Introduction

The Shampay House project of 1919 has been commonly understood as one of Frank Lloyd Wright's cruciform Prairie houses. It was planned to be erected in Beverly Hills, Illinois, but never realized. The demise of the project might have been due to financial difficulties. It has been assumed that Wright designed the house in his Tokyo office, and that Rudolph Michael Schindler altered it in the U.S. However, at the outset, Schindler claims authorship of the work. In a previous paper, Park and March [2002] proved that the authorship of the Shampay house design is attributable to Schindler based on newly-found correspondence and drawings.

The earlier historic discussion left out most of the stylistic characteristics of the project. Schindler's stylistic tendencies characterize his architectural practices both before and shortly after the Shampay project. This paper examines in detail the stylistic distinctions between Wright and Schindler's work at that time. First, there is the general spatial arrangement of the scheme; second, the use of symmetry; and third, proportional design. Some commentators have speculated about the characteristics of the project, but their conclusions have remained unexplained. For example, David Gebhard gives Schindler negligible credit for his contributions to the design: in particular, the placement of the garage as part of the main body of the house [1980, 37-38]. B. Giella differentiates its characteristic features from Wright's typical Prairie designs, describing the "asymmetric" spatial configuration, "the placement of the bank of tall, narrow windows" and "the blank wall" [Giella 1985, 243-246]. Futagawa and Pfeiffer claim the Shampay house as a direct source for the development of Wright's Usonian houses fourteen years later [Futagawa and Pfeiffer 1985, 192]. These historians devote their research to the characteristics of the design. However, many of their arguments have risen partly through misunderstanding and misinterpretation of the project. Hence, detailed examination of several key stylistic features of the project is required to identify Schindler's contribution to the project, thereby confirming the authorship.

2 Spatial arrangement

Wright acknowledges that Schindler has captured the Prairie house style. In their paper "The language of the prairie: Frank Lloyd Wright's prairie houses," architects H. Koning and J. Eizenberg make use of Wright's own descriptions and buildings to define a formal grammar [1981]. "Consistency in grammar is therefore the property – solely – of a well-developed artist-architect," Wright wrote towards the end of his life [Wright 1954, 182].

It appears that Schindler had familiarized himself with the grammar of the Prairie before coming to the States, starting with the drawings in the Wasmuth portfolio and then by visiting and photographing Wright's houses in the Chicago area.[1] He would also have been familiar with the latest Taliesin practices: "one of the best of my houses," as Wright described, designed for the future Governor Allen of Kansas, was barely off the drawing board when Schindler joined, and was awaiting Paul Mueller's release from his Imperial Hotel commitment before construction

1590-5896/06/010017-16 DOI 10.1007/s00004-006-0002-1
© 2006 Kim Williams Books, Firenze

commenced [Futagawa and Pfeiffer 1985, 126]. Koning and Eizenberg quote Wright on the underlying principles of the Prairie house: first, the fire, the hearth is at the center of the home, the chimney should be a "broad generous one... kept low on gently sloping roofs, or perhaps flat roofs" [Wright 1953, 136-137]; second, the simplification and consolidation of functional areas on the principal floor by elimination of partitions and doors – a "more 'free' space and more livable too" – and "cutting off the kitchen" with other service spaces "semi-detached" [Wright 1953, 139]; third, to reduce throughout the dwelling the necessary parts of the house and the separate rooms to a minimum; fourth, to associate the building as a whole with the site by extension; fifth, "to eliminate the room as a box and the house as another" [Wright 1953, 14].

The Shampay achieves all of these criteria. Wright's concern about the size of the chimney is consistent with his views, but it is arguable whether Schindler had not already made the chimney "broad and generous." The dining room, living room and entry lobby all flow together spatially without partitions or doors, with the kitchen and maids' room "semi-detached" from the living area behind the hearth. It follows that separate rooms are kept to a minimum, and even on the second floor the arrangement is remarkably simple. The house is locked into its site with a well-developed parterre. This is particularly evident in the August foundation plan.[2] The house exhibits considerable transparency, especially in the see-through living room, the dining area with its conservatory glazed on three sides, and the entry area and porch. On the second floor, the owner's room shows similar openness. In a Wrightian manner, Schindler pries the corners apart and lets the roofs appear to float above ranges of fenestration – the box is destroyed. The Shampay is in this sense a classic, even conservative, Prairie house. Yet Pfeiffer claims "elements in this Shampay house point directly to the Usonian houses that come fourteen years later" [Futagawa and Pfeiffer 1985, 192].

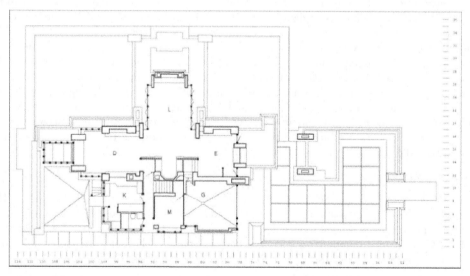

Fig. 1a. The Shampay House, First Floor Plan, Reconstruction from June 9, 1919 drawings: ordered numbers on each unit in the first floor plan identifies the division of 2-foot unit grid. [G: garage, L: living room, E: Entry, D: dinning room, M: maid room, and K: kitchen]

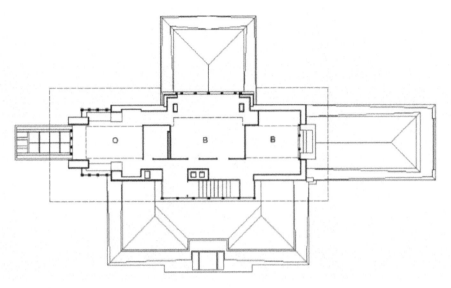

Fig. 1b. The Shampay House, Second Floor Plan, Reconstruction from June 9, 1919 drawings [O: owner's room, B: bed room]

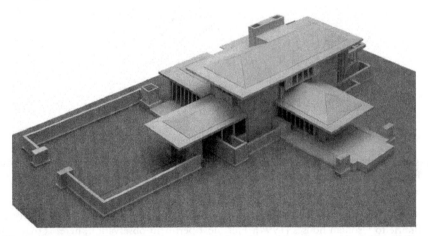

Fig. 1c. The Shampay House, 1919, a quarter inch scale model constructed by Adrian Shih

Wright gives nine points that identify the Usonian: first, visible roofs are unnecessary; second, garages are not required, a carport will do; third, basements are not required; fourth, interior "trim" can be eliminated; fifth, no radiators, no light fixtures; sixth, furniture and pictures can be built-in; seventh, no painting of surfaces; eighth, no plastering of walls; ninth, no gutters, no downspouts.[3] The roofs are very visible in the Shampay design, and Wright even wanted to enhance that visibility. The house has a built-in garage. The house has a large basement area. None of these satisfy Wright's specification for a Usonian. The other requirements concern details and finishes, some of which are true of Prairie houses as well.[4]

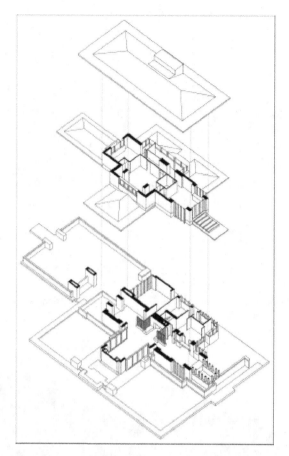

Fig. 2. The Shampay House (1919), Exploded Axonometric

What is absolutely certain is that Schindler, just two years after the Shampay project, designed and built his own house, Kings Road House (1921-22), in which all nine of these Usonian criteria were radically met.[5]

Terry Knight has also considered the stylistic transformations required for the development from the Prairie to Usonian houses [1994]. Her most pertinent observation, here, is that typically the hearth is positioned differently in the two types. Formal, grammatical details aside, the fireplace in the Usonian is placed in the single space of the living area in such a way as to define distinct functional zones, particularly the entry and areas for sitting, dining and cooking without the need for partitions and doorways. In the Prairie house the fireplace usually just marks one end of the living area. Again the Shampay conforms not at all to the typical Usonian configuration, but falls nicely into the Prairie arrangement.[6]

3 Symmetry

Not only would Wright have surely known of Viollet-le Duc's comments on symmetry, [Viollet-le-Duc 1959, 274-275] but also Wright himself had written in 1908:

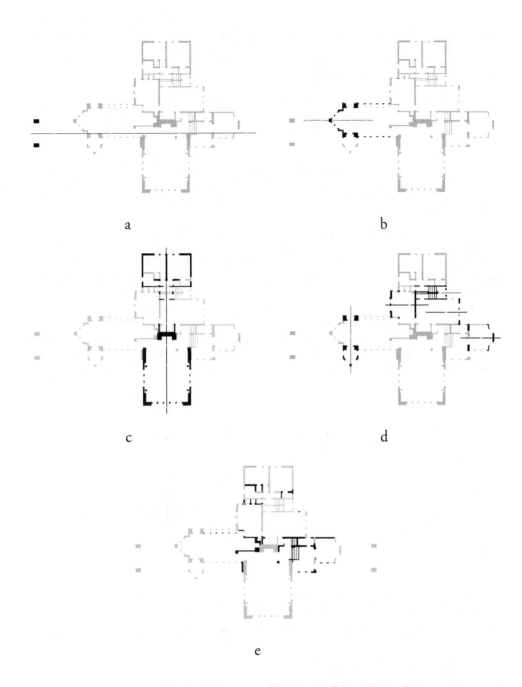

Fig. 3. Subsymmetry drawings for the first floor plans of the Ward Willits House, 1901: a. long axis through porch piers; b. axis through the dining room; c. axis through the living room and hearth; d. minor axes through various parts of the plan; e. remaining elements having no overall symmetry

In laying out the ground plans for even the more significant of these buildings a simple axial law and order and the ordered spacing upon a system of certain structural units definitely established for each structure in accord with its scheme of practical construction … and although the symmetry may not be obvious always the balance is usually maintained [Wright 1941, 39].

Wright's Prairie principles of spatial composition resonate with his tenet of symmetry.

It is reasonable to assume that Schindler – who was an admirer of Wright and who respected Wright's philosophies of design – was also aware of Otto Wagner's view[7]:

A simple, clear plan in most cases requires the symmetry of the work. In a symmetrical arrangement there is some measure of self-containment, completeness, balance; an impossibility of enlargement; even self-assurance … The aping of unsymmetrical buildings or the intentional making of an unsymmetrical composition in order to achieve a supposed painterly effect is totally objectionable [Wagner 1988, 86].

Wright made frequent use of global, axial symmetry in his public works, including the Imperial Hotel. In his domestic work up to 1919 such obedience to symmetry is rare although the more formal spaces often exhibit local symmetry. In the 1909 Como Orchard Summer Colony, illustrated in the Wasmuth portfolio, the clubhouse and many of the cottages are axially disposed along the crest of the hill, but the symmetry is broken where the terrain makes it impossible to maintain, or where the variety of accommodation is required [Wright 1910, pl. 100]. This is a good illustration of Wright's approach: symmetry is used where it makes some sense, and not pursued elsewhere. The most Beaux Arts of all of Wright's domestic designs is the unyielding axial project for the C. Thaxter Shaw mansion in Canada.[8] An elaborate play of multiple axiality is to be seen in the 1901 Willits house [Wright 1910, pl. 70].

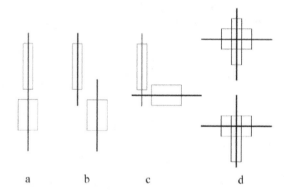

a   b   c   d

Fig. 4. Classification of axialities in residential designs from the Wasmuth portfolio: a. plain axial; b. staggered axiality; c. angled axiality; d. cross, or T, axiality.

Four types of axiality may be discerned among residential designs in the Wasmuth portfolio:

- one simple axis;
- two parallel axes, or staggered axiality;
- two non-intersecting axes at right-angles, or hinged axiality;
- two orthogonal axes forming either a T, or cross depending on the degree of overlap (fig. 4).

The latter type of axiality is to be found in the Shampay. Schindler, in fact, turns the clock back.

A survey of built projects by Wright in the five years before the Shampay house design shows little use of symmetry in house planning, but a residual use appears in the principal façades. Wright's remarkable Bach residence (1915) exhibits bilateral symmetry to the street, preserves much symmetry to each side, and abandons it altogether at the rear; in the interior the living space is freely and asymmetrically planned. The American System-Built Homes that were built from 1915 mostly eschew axiality.[9]

The fine Allen residence designed by Wright in 1916 has some vestigial axiality in the living room only, and this is staggered: one axis can be drawn through the fireplace, and another parallel axis through the opposing window wall. The axes run some three feet apart. Wright's Japanese residences (1917-1918) possess only rare instances of local symmetry. The Olive Hill project, which Schindler supervised, hints at axiality and symmetry for the major spaces in the three residences, but on each occasion strictness is deliberately frustrated by the intrusion of some local incident, such as the location of the fireplace or – in the case of residence B – the accommodation of a dining area [Smith 1992]. His practice shows that Wright himself was largely finished with axiality and symmetry as major form-makers in residential buildings by and during Schindler's tenure.

What of Schindler? His Wagnerschule student projects mostly adhere to strict axial symmetry,[10] although Schindler's early sketches indicate an interest in picturesque groupings in vernacular architecture [March and Scheine 1994, 51]. Schindler's 1914 submission for a neighborhood center in Chicago holds to strict bilateral axes.[11] Schindler's adobe proposal for the Thomas Paul Martin residence in Taos, New Mexico, is an academic study in bilateral symmetry and very much in tune with Wagnerschule student schemes [E. Smith et al. 2001, 186, 274].

An axial plan for a community center in Wenatchee, prepared by Schindler in May 1919, was criticized by Wright in the same letter of 25 June in which he complimented Schindler on the Shampay design. In a sketch, Wright turns the building at right angles to the "axial street," which enters the park on its long side, to create a new axis down the center of the park and parallel to the Columbia River.[12] Both the Monolithic house designed by Schindler in Wright's absence [K. Smith 1992, 104-115], and another independent project for a residence in Oak Park, dated c1917, are somewhat stiffly axial in plan, especially the latter [E. Smith et al, 2001, 185-186, 187-188].

Schindler's 1920 competition entry for a branch of the Free Public Library, Jersey City, New Jersey, demonstrates his extraordinary skill in orchestrating distinct symmetries, a facility to be continued in other early independent works such as the 1922 Popenoe Cabin, the 1925 How house, and through to the 1930s with the Schindler Shelters [Park 1996, 2000, 2001]. The Buena Shore Clubhouse, designed by Schindler for Ottenheimer, Stern and Reichert before joining Wright, is an asymmetrical L-form with local axialities in the principal rooms [Giella 1994]. The 1918 Log house designed at Taliesin belongs to more progressive developments in Europe influenced by Gottfried Semper's *Der Stil* and the transmission of these ideas especially in Henrik Berlage's works and writings – the Dutch artist Vontongerloo comes to mind – than anything Schindler might have learned from Wright.

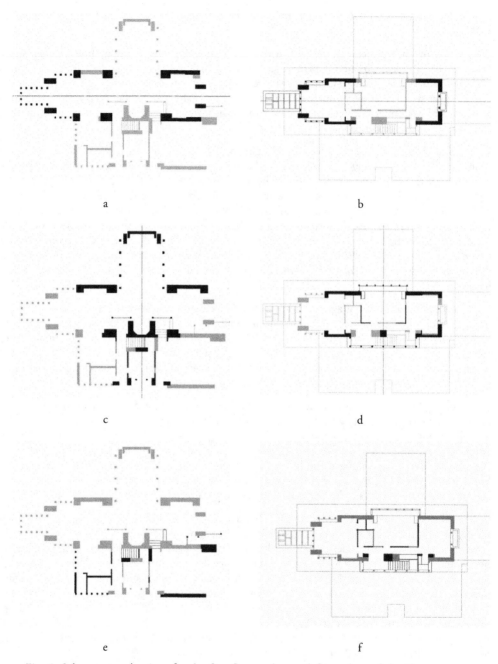

Fig. 5. Subsymmetry drawings for the first floor and second floor plans of the Shampay house: a. and b. elements and sub-elements, which are bilaterally symmetrical along the longitudinal axis; c. and d. elements and sub-elements along the latitudinal axis; and e. and f. the remaining elements having no overall symmetry

Axiality is buried in just one of the layers of the log sub-structure, and bilateral symmetry is found only at certain cross-sections; otherwise the plans and elevations of the Log Cabin are studies in asymmetrical design [March 1994b].

The Schindler house of 1921-22 is without symmetry in any conventional sense. The Shampay house falls under that class of Prairie houses with intersecting, orthogonal axes.

An analysis of the Shampay house plans (fig. 5), clearly indicates that much of the design is accounted for by two overlapping parts, each of which is bilaterally symmetrical. The symmetrical parts are assigned to the service areas. One consequence of the intersection is that global symmetry is lost even while local symmetries are preserved. Schindler uses a similar approach in the basement and first floor plans of the 1920 Library project where the auditorium, children's and reading rooms are located [Park 1996, 80, fig. 14].

4 Proportion

By the time the Shampay house was designed, Schindler was engaged in a search for a reliable unit system that he could invariably apply to composition as well as construction. The Shampay house plans are set out over a square grid (see fig. 1a). The June plans are shown with a 2' x 2' grid with every other gridline marked at 4' intervals along the bottom and up the right-hand side of the drawing. These plans were drawn at a scale of 1/4" to 1', so that the grid lines were set 1/2" apart. The blueprints sent to Wright in May indicate a 4' x 4' grid. Since the May blueprints were scaled 1/8" to 1' foot, these gridlines are also shown at 1/2" apart.

Wright did not usually place gridlines on house plans before the concrete block houses of 1923. One example is the Stewart house. This house had been left to van Holst to "detail" when Wright left for Europe in 1909 [Alofsin 1993, 312]. One of the earliest house designs to show a grid on the plan is the George Gerts Double house, 1902, where mostly a 3' x 3' square unit is used [Storrer 1993, 74, S.077, T.0202]. The same unit is employed throughout the Mrs. Thomas H Gale Cottage, 1909 [Storrer 1993, 84, S.088.1, T.0521]. The buildings for the Bitter Root Inn, 1908, and the Como Orchard Summer Colony, 1909, used a 3'-6" square unit shown on plans [Storrer 1993, 146-8, S.144, T.1002, S.145, T.0918A]. The American System-Built Homes are set out on a 2' x 2' square unit system, but plans do not show the grid.[13] The three summer houses at Grand Beach, Michigan, 1916, mostly work to a 2' planning module [Storrer 1993, 200-1, S.196, T.1601; S.197; T.1607; S.199, T.1603]. Three more houses are attributed to Wright in Japan: two use a double tatami module, 6' x 6', and one a 4' x 4' unit system [Storrer 1993, 209, S.206, T.1702; 210, S207, T.2901; 215, S.212, T.1803]. The Olive Hill development is shown with a 20' x 20' grid derived from the original olive tree planting, which is further subdivided into 4' square units for architectural planning purposes. This 4' module is seen extended upwards in the elevations, but it is not used to determine vertical dimensions [K. Smith 1992].

Not all of Wright's houses of this same period use a unit system. The plans for the Emil Bach house (1915) show many different dimensions which cannot simply be related to a regular grid [Futagawa and Pfeiffer 1985, 63, figs. 88, 89]. The Sherman Booth house of the same year shows one dimensional scheme for the living room and a different one for the dining room [Futagawa and Pfeiffer 1985, 66, fig. 94]. The symmetrical façade to the Frederick C Bogk house, 1916, divides into alternating solids and voids in a subtle play on the brick length [Futagawa and Pfeiffer 1985, 108-109, figs. 190-192]. Like the Booth house, Governor Allen's residence (1916) uses different and distinctive dimensional sets for separate parts of the house: the living room, the dining room, kitchen, the entry and the porte cochere [Wright 1994, 35].

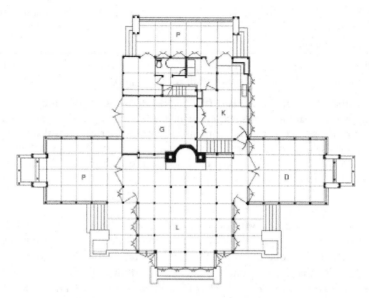

Fig. 6. F.L. Wright, the Stewart House, Montecito, CA, 1909. The garage is built on the main floor of the house next to the chauffeur's room in the original design. Later on, the design was changed and built while Wright was in Europe from late 1909 to early 1911 (Storrer, 1993, 162-3). In the final design, the servant's quarters including the kitchen, pantry, maid/chauffeur, laundry, and bath are enlarged by removing the garage. The functional arrangement of the Shampay house is almost identical to the original design of the Stewart house Key: G, garage, L, living room, P, porch, D, dining room, and K, kitchen

Schindler had used a 6'-6" square unit in laying out the plans for the Buena Shore Club (1917) before joining Wright [Giella 1994, 42]. While at Taliesin, he produced drawings for the Log Cabin on a 2' square unit which is presented on the two principal plans [March 1994a]]. Unlike Wright's work up to this time, the square unit system also controls the dimensions in elevation. This is a simple matter since the logs are either 8" or 6" square in section – three or four logs fit into each 24" module. But what makes the Log Cabin scheme stand out is Schindler's first known use of his space reference frame [March 1994a].

Here the grid is marked along the bottom with numbers from 1 to 26, and up the right-hand side with letters A to Q. This is original to Schindler. Schindler's 1920 competition entry for the Free Public Library shows neither, although, on a site 100' x 100', the building is planned within nine squares 20' x 20', each further sub-divided into 4' square units.[14] The Shampay house shows a grid, but does not use the reference frame.

Whereas the main outlines of the Shampay house plan relate to the unit system, detailed considerations give rise to divisions such as 7"+ 41" + 7", the mullion + French door + mullion pattern of the east and west living room walls, or the 7" + 17" + 7" pattern of the dining room walls (fig. 7). From Charles E. White Jr., apprentice for Wright, we know of Wright's use of unit system employed in "the casement window unit of about these proportions" [Nancy and Smith 1971]. While this is not the place to go into technicalities, the first division is, center to center of the mullions to the opening, in the ratio of √4:√3,[15] and the second in the ratio of √2:1. These are typical relationships to be found in Wright's early work.[16] For example, the living room piers and openings in the Allen house have the pattern 16" + 68" + 16" which gives 84" center to center between piers: a proportion of 84:68::21:17, or $\sqrt{3}:\sqrt{2}$.[17] It is not necessary to assume that Schindler was influenced by Wright in this. Otto Wagner had been a master of proportional design, and Schindler, who was known to be very adept at numbers, had excelled in the Wagnerschule ornamental design class – essentially a course in applied geometry. He would have appreciated and been familiar with the game that Wright played. As Wright once wrote to Schindler concerning a personal matter, "your instinct for proportion and a few other things may help save some of the pieces."[18]

Fig. 7. Photograph of the Shampay house model showing the living room and dining room fenestration patterns

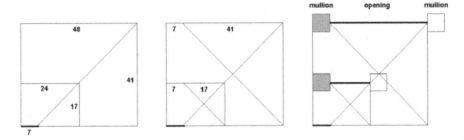

Fig. 8. R. M. Schindler's row-7 construction of wall openings in the Shampay house. Left, a length 7 is drawn and a 45° line is struck from the end of the line. All rectangles of integer length which lie along the 45° line belong to row-7. Center, counter 45° lines cut the lengths of the 17/ 24 and 41/ 48 rectangles into the dimensions found in the Shampay French door and window plans. Right, namely the 7 + 41 + 7 and 7 + 17 + 7 patterns

Despite the similarities with Wright's proportional schemes, the living room and dining room openings in the Shampay design provide perfect proof of Schindler's own preoccupation with a special kind of proportional theory at this time, which might be described as his unique signature. Two years before the Shampay project was on the drawing board, Schindler had delivered a series of lectures at the Emma Church School in Chicago. When he came to discussing proportion, he rejected the principle of the regulating line (which was favored by his European contemporary Le Corbusier), in favor of a theory of "rows."[19] It is a proportional theory which is completely compatible with Schindler's use of a modular planning grid. Schindler used the term "row," a sequence, "a following of unequal units with definite changes". If a/b is a term in the "row", the next term is (a + 1)/(b + 1). In these Shampay openings, Schindler employs two elements from the seventh row, row-7: 1/8, 2/9, 3/10, ... 17/24, ... 41/48, 42/49, ... (fig. 8).

It will be noticed that the common difference between numerator and denominator is 7, hence row-7. Now, 24/17 is a good rational approximation to √2, the diagonal of a unit square; while 41/48 is very close to 42/49 = 6/ 7, where 7: 6 is a well known rational convergent for √4:√3, the side of a unit equilateral triangle to its altitude (fig. 9). And so the use of the "row" system allows Schindler to choose elements which relate to geometrical constructions, and thence to the drafting squares and the tiles of the Seventh Froebel Gift (fig. 10).

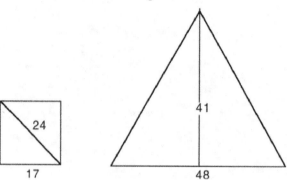

Fig. 9. The rational ratios 24:17 and 48:41 approximate the diagonal of a square (24) to its side length (17), and the side of an equilateral triangle (48) to its altitude (41)

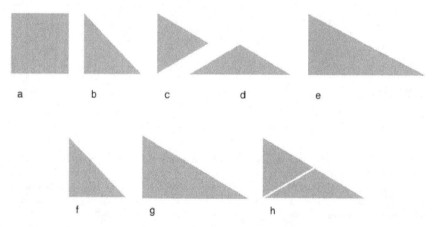

Fig. 10. Top row: Froebel's tiles from the Seventh Gift. a. a square; b. half square, or right isosceles triangle; c. equilateral triangle; d. 120° obtuse isosceles triangle; e. half of a double square, a right scalene triangle. Bottom row: drafting triangles; f. 45° triangle; g. 30°, 60° triangle; h. 30°, 60° triangle composed of Froebel tiles c and d

Fig. 11. Fenestration proportions in the Shampay house compared with drawing board set squares and tiles from Froebel's Seventh Gift. a. the 45° set square, or Froebel's half-square tile, proportion the spacing of mullions and openings in the dining room; b. Froebel's obtuse isosceles triangle and the half double-square tiles determine the spacing of mullions and french doors in the living room; c. 30°, 60° drafting triangles in two orientations give rise to the same spacing

5 Conclusion

The authorship of the Shampay house has long been a subject of speculation. This formal analysis of the project confirms our earlier argument that while Wright has the ownership of the drawings, Schindler has the sole authorship of the work. Moreover, Schindler deserves full credit for the stylistic characteristics of the project.

The detailed analysis proves that the Shampay house is a conservatively designed Prairie house in Wright's grammar, with no embryonic Usonian features. The claim by Bruce Brooks Pfeiffer that the Shampay looks forward to the Usonian houses of fourteen years later would make Schindler in some part responsible for that development through this design.

At age 34 Wright designed and built the Willits house (1901), which exhibits a masterly command of multi-axial planning and modular flexibility. At age 32 Schindler designed the Shampay house in Wright's name. Apart from the service areas, the house is rigidly biaxial and strictly modular, based on the 2′ unit square. The scheme is a dutiful "Lloyd-Wright facture;" like the Wenatchee project, it could be described as "a little too 'old school'." Nothing here indicates the radical form that Schindler's own house was to take two years later which truly set the agenda for the future Usonian houses.

Notes

1. For example, there is a 1914 letter dictated by Wright, in the year of Schindler's arrival in Chicago, introducing him to Mrs. Coonley so that he might see her house. Several of Schindler's photographs of Wright's works are reproduced in [March and Sheine 1994, 50].

2. There are three different institutions that own Shampay house drawings. Two complete sets of drawings are available from the Getty Museum, copyright under the Frank Lloyd Wright Foundation. One set of drawing is dated 9 June 1919, Getty 1903.002-011, and the other set is dated from 13, 18 to 20 August 1919, Getty 1903.012-017. Both sets are signed 'Frank Lloyd Wright Architect per R M Schindler [signed]'. Among the drawings, there is an early schematic drawing, which does not belong to either set, 1903.001. It includes the second and basement floor plans, dated May 1919. In the Architectural Drawing Collection at the University of California, Santa Barbara, there are some schematic sketches, several blueprints and two perspectives, Garland 3234-3246 [Gebhard 1993, Drawing no. 3234-3246]. Recently, Maureen Mary discovered a set of six previously unknown and recently found blueprints. In 1919, Schindler worked on the design of the Shampay House from April 15 to May 12 when he turns to a commission for a community center in Wenatchee. Six 1/8″ to 1′ scale blueprints of the project are sent to Wright in Tokyo at this time, one of which, the first floor plan, is dated 4 May 1919. Penciled amendments on the Shampay House blueprints were mailed from Tokyo by Wright, postmarked June 4, and addressed: to 'Rudolph Schindler, c/o Frank Lloyd Wright, 1600 Monroe Bldg., Chicago, Ill., USA'. The envelope is marked in pencil 'Rec. June 24th 1919'.

3. [Wright 1954, 82-83]. The full text is on its side accompanied by two preliminary sketches of the prototypical Jacobs Usonian of 1937.

4. The Wasmuth portfolio refers to the body of work published by Ernst Wasmuth in 1910 [Wright 1910] entitled *Ausgeführte Bauten und Entwürfe*: "It is still in a comparatively primitive stage of development; yet radiators have disappeared, lighting fixtures are incorporated ... disfiguring down-spouts ... become useless in the winter." Wright also indicates that he would like to see less furniture "at large" in his plans, and no easel pictures on the walls, but built-in artworks. He had already started to minimize the plastering of walls and the use of paint, often preferring exposed brick surfaces and natural wood.

5. [K. Smith 1987, 26]. Smith enumerates several other features of the typical Usonian foreshadowed by this house. See also [K. Smith 2001].

6. Nor does the Shampay conform as a Usonian to Storrer's diagram of the "Transformation from Prairie to Usonia." See [Storrer 1993, 219].

7. Wagner was taught under a former assistant of Karl Schinkel, C F. Busse at the Königliche Bauakademie in Berlin (1860-61) [Wagner 1988; Giella 1985]. Also, Wagner's theoretical development of modern architecture was deeply rooted in Semper's theory, including not only practical aspects, such as material and technology, but also formal compositional principles. Wagner's Beaux-Arts acquaintance was from the Academy of Fine Arts under August von Siccardsburg and Edward van der Nüll [Wagner 1988, 86].

8. The dramatic poché is [Wright 1910, pl. 70].

9. [Storrer 1993, S203 1-4, S204 1-6]. The Heisen residence S204.6 is a possible exception.

10. For example: the elevations to the undated neo-classical Post Office [March and Sheine 1994, 46]; and to his 1912-13 thesis project, "A Crematorium for a City of Five Million" [E. Smith et al. 2001, 17, 234].

11. [E. Smith et al. 2001, 21 and 186]. Wright's contemporary hors concurs submission for a Chicago quarter-section is remarkably free from Beaux Arts influences.

12. Getty Box 1 Folder 10. Wright comments: "Your layout of Wenatchee at hand. If anything comes of it – turn the main group of buildings side on the main axis – letting the approach from the city through to the river with a raking view of the facades. It is now a little too 'good school' – a pompous 'obstruction' rather than a sympathetic auxiliary to the natural features of the site. Suggest like this (sketch)."

13.   It is entirely possible that the plans were drawn on paper placed over an underlying sheet showing a grid. Such a sheet, with 1″ squares, was found among some Schindler drawings, for example.

14.   The same large and smaller divisions used in the Olive Hill site plans [Park 1996].

15.   The rational ratio here is 48:41. 71:41 is a convergent to $\sqrt{3}$:1. Since $3/\sqrt{3}=\sqrt{3}$, it follows that 41:71/3 is also a convergent to $\sqrt{3}$:1. But 71/3 is approximately 24. Thus, 41:24 can be said to be a near convergent to $\sqrt{3}$:1. 41:48 is thus $\sqrt{3}$:2, or 48:41::$\sqrt{4}$:$\sqrt{3}$.

16.   Ratios easily derived, using arithmetical approximations, from the 7th Froebel gift, as well as simple relationships taken from the diagonals of the unit cube [March 1995].

17.   The well known rational convergents 7:4 for $\sqrt{3}$:1, and 17:12 for $\sqrt{2}$ give this result.

18.   Getty Box 1, Folder 16, letter postmarked Tokyo, April 1920.

19.   This theory is fully explained in [March 2003] and [Park 2003].

References

ALOFSIN, A. 1993. *Frank Lloyd Wright The Lost Years, 1910-192: A Study of Influence*. Chicago: The University of Chicago Press.

FUTAGAWA, Y. and B.B. PFEIFFER. 1985. *Frank Lloyd Wright Monograph 1914-1923*. Tokyo: A.D.A. Edita.

GEBHARD, D. 1980. *RM Schindler*. Santa Barbara and Salt Lake City: Peregrine Smith.

———, ed. 1993. *The Architectural Drawings of R.M. Schindler*. Garland Architectural Archives. The Architectural Drawing Collection, University Art Museum, University of California, Santa Barbara. New York: Garland Pub.

GIELLA, B. 1985. R. M. Schindler's Thirties Style: Its Character (1931-1937) and International Sources (1906-1937). New York: Ph.D. Dissertation, Columbia University.

———. 1994. Buena Shore Club. Pp. 38-47 in *R.M. Schindler: Composition and Construction*, L. March and J. Sheine, eds. London: Academy Editions.

KNIGHT, T.W. 1994. *Transformations in Design: a Formal Approach to Stylistic Change and Innovation in the Arts*. Cambridge: Cambridge University Press.

KONING, H. and J. EIZENBERG. 1981. The Language of the Prairie: Frank Lloyd Wright's Prairie Houses. *Environment and Planning B* **8**:295-324.

MARCH, L. 1994a. Proportion is an Alive and Expressive Tool. Pp. 88-101 in *R.M. Schindler: Composition and Construction*, L. March and J. Sheine, eds. London: Academy Editions.

———. 1994b. Log House, Urhutte and Temple. Pp. 103-113 in *R.M. Schindler: Composition and Construction*, L. March and J. Sheine, eds. London: Academy Editions.

———. 1995. Sources of characteristic spatial relations in Frank Lloyd Wright's decorative designs. Pp. 12-49 in *Frank Lloyd Wright: The Phoenix Papers*, L.N. Johnson, ed. Tempe: Herberger Center of Design Excellence, Arizona State University.

———. 2003. Rudolph M. Schindler: Space Reference Frame. Modular Coordination and the Row. *Nexus Network Journal* **5**:48-59.

MARCH, L. and J. SHEINE, eds. 1994. *R.M. Schindler: Composition and Construction*, London: Academy Editions.

NANCY, K. and M. SMITH, eds. 1971. Letters, 1903-1906, by Charles E. White Jr., from the Studio of Frank Lloyd Wright, *Journal of Architectural Education* **25**:104-112.

PARK, J-H. 1996. Schindler, Symmetry and the Free Public Library, 1920, *Architectural Research Quarterly* **2**:72-83.

———. 2000. Subsymmetry analysis of architectural designs: some examples. *Planning and Design* 27:121-136.

———. 2001. Analysis and Synthesis in Architectural Designs: A Study in Symmetry, *Nexus Network Journal* **3**:85-97.

———. 2003. Rudolph M. Schindler: Proportion, Scale and the 'Row'. *Nexus Network Journal* **5**:60-72.

PARK, J-H and L. MARCH. 2002. The Shampay House: Authorship and Ownership. *Journal of the Society of Architectural Historians* **61**:470-479.

SMITH, E., et al. 2001. *The Architecture of R.M. Schindler*. Los Angeles: The Museum of Contemporary Art.

SMITH, K. 1992. *Frank Lloyd Wright: Hollyhock House and Olive Hill*. New York: Rizzoli.

———. 1987. *R M Schindler House, 1921-22.* West Hollywood: Friends of the Schindler House.

———. 2001. *The Schindler House.* New York: Henry N. Abrams.

STORRER, W.A. 1993. *The Frank Lloyd Wright Companion.* Chicago and London: The University of Chicago Press.

VIOLLET-LE-DUC. 1959. *Discourses on Architecture II.* Benjamin Bucknall, trans. New York: Grove Press.

WAGNER, O. 1988. *Modern Architecture: a Guidebook for his Students to this Field of Art.* H. F. Mallgrave, trans. Santa Monica: The Getty Center

WRIGHT, F.L. 1910. *Ausgeführte Bauten und Entwürfe.* Berlin: Ernst Wasmuth. http://www.lib.utah.edu/digital/wright/.

———. 1941. 1908: In the Cause of Architecture, I. In *On Architecture – Selected Writing 1894-1940.* New York: Duell, Sloan and Pearce.

———. 1953. *The Future of Architecture.* New York: Horizon Press.

———. 1954. *The Natural House.* New York: Horizon Press.

———. 1963. *Frank Lloyd Wright Buildings Plans and Designs.* New York: Horizon Press.

———, et al. 1994. *Frank Lloyd Wright: The Complete Wendigen Series.* New York: Gramercy Books. Reprint of *The Life Work of the American Architect Frank Lloyd Wright,* Wendigen, 1925.

About the author

Jin-Ho Park Ph.D. currently teaches architectural design, theory and history in the Department of Architecture at Inha University, Korea. Prior to joining Inha University, Dr. Park was a tenured associate professor in the School of Architecture at the University of Hawaii at Manoa. He earned his BS in architecture from Inha University, Korea, and his MA and Ph.D. in architecture from UCLA. He is the first recipient of the R.M. Schindler Fellowship of the Beata Inaya Trust Fund, 1996/7, and twice recipient of the UCLA Chancellor's Dissertation Fellowship, 1998/99. He was awarded the University of Hawaii Board of Regent's Medal for Excellence in Teaching in 2002 and the ACSA/AIAS New Faculty Teaching Award in 2003. In 2005, he received the Journal of Asian Architecture and Building Engineering (JAABE) Best Paper Award. The focuses of his academic research are on Architectural History and Theory, Fundamentals of Architectonics: Proportion, Symmetry, and Compartition, The Architecture of R. M. Schindler, Design Computation, Shape Grammars and Digital Media. He has been publishing in various refereed scholarly journals including *Architectural Research Quarterly (arq), Environment and Planning B: Planning and Design, Journal of the Society of Architectural Historians, Journal of Architectural and Planning Research, Journal of Architectural Education,* the *Mathematical Intelligencer, Journal of Asian Architecture and Building Engineering, The Journal of Architecture, Open House International,* and the *Nexus Network Journal.*

Steven Fleming

C/- Romberg Building
University of Newcastle
Callaghan, NSW, 2308, AUSTRALIA
Steven.fleming@newcastle.edu.au

Mark Reynolds

667 Miller Avenue
Mill Valley, CA 94941 USA
marart@pacbell.net

Timely Timelessness: Traditional Proportions and Modern Practice in Kahn's Kimbell Museum

The twentieth century witnessed declining interest in architectural proportioning systems, which were virtually eclipsed by technical, social and fiscal agendas. Louis Kahn is a seminal architect, whose most acclaimed building, the Kimbell Art Museum (1966-72), represents a compelling case-study in the use proportions by twentieth-century architects. In spite of a raft of peculiarly modern restrictions (both technological and programmatic), Kahn appears – despite his espoused ambivalence concerning proportion – to have intentionally produced a building with an array of approximate geometrical as well as precise harmonic proportions.

This two-part paper presents the findings of a multifaceted research project that examined the Kimbell's proportions from numerous standpoints. Part 1 presents a textural analysis of Kahn's statements regarding proportion, as well as the findings of an archival study of correspondence between the architect and his client and consultants. Part 2 presents a *prima facie* geometrical analysis of the construction drawings for the project. The division into parts reflects an apparent discrepancy between Kahn's buildings and what he had to say about them.

PART 1

Introduction

If the proportions of buildings are symbolic, then in the case of much twentieth-century architecture, mathematics symbolises the cultural authority of nothing more than everyday instrumentalism, the overwhelming imperative to produce buildings on time and on budget for profit-orientated clients. Architects often make a virtue of standard-sized products – such as sheets of form-ply – thus producing symbols of industrial production. Design decisions often fall to others on the design team. Structural engineers, for example, may determine a space's width according to the maximum span of a commonly available steel section. A services engineer can reduce the overall height of an office block simply by specifying slimmer ductwork for its ceilings. When contemporary architects bring out their calculators, it is not usually to find the golden mean, but to calculate permissible floor space ratios, or fire egress limits, or the number of toilets required on a floor. The proportions of buildings are often subservient to, and symbolic of, a plethora of apparently mundane concerns.

But is this true of all modern buildings? Perhaps among a rarefied minority of great works of architecture there are cases in which the noble tradition of inscribing meaningful proportions has been kept alive by architects dedicated to the art, rather than the business, of architecture. With this question in mind, the present paper examines the most revered building of the celebrated modern master, Louis I. Kahn, whose name is often invoked as a synonym for devotion to the art of architecture. If any twentieth-century architect ever resisted the everyday obstacles that fetter most architects' symbolic geometrising, it was Kahn. The paper will focus on the Kimbell Art

Museum in Fort Worth, Texas (1966-72), and approach the topic from a number of angles. Firstly, Kahn's public statements concerning proportion will be considered. Secondly, geometrical analyses of the plans and sections of the Kimbell are presented, showing where precise and approximate proportions actually exist. Analytical mathematical and geometrical findings are then reconciled with the archival records for this project in an attempt to determine the extent to which existing proportions reflect deliberate architectural intentions.

The paper highlights a complex interplay of forces and draws few general conclusions. For Kahn, the incorporation of harmonic proportions (that is, proportions based on musical ratios) is a stated aim, but it is not an overriding concern. Part of his genius could rest in his ability to frame correspondence with his client and specialist consultants in terms that would make music-based proportions a likely outcome of their dialogue. In particular, Kahn couched queries and suggestions regarding dimensions in terms of whole feet dimensions. When dealing with major dimensions, he errs away from prime numbers, preferring numbers with factors that could be shared with related dimensions. In this way, he is able to appease others' instrumentalist and fiscal agendas while leaving himself greater scope for establishing music-based ratios. The apparent use of geometric proportions in the plan suggests another step in Kahn's design process, in which the plan was massaged until it complied with classic geometric grids. The identification of these grids, which appear to underpin the plan, is the most fascinating aspect of this inquiry, as the archival and oral histories provide no trace of such a step in the Kimbell's design or documentation phases.

The complex and tentative nature of these findings reflects the disparate backgrounds of this paper's authors. Mark Reynolds is a geometer whose publications in this field are specifically concerned with finding geometrical relationships. Steven Fleming is an architectural historian who has published texts challenging geometrical analyses, showing their inconsistency with archival records. The present work is the result of a genuinely dialectical collaboration, involving a meeting in Mexico, exchanged plans, and over 300 emails by two scholars who have relentlessly scrutinised one another's positions, culminating in a joint publication effort. The hope now is that readers of the *Nexus Network Journal* will engage in this debate, leading, ultimately, to a better picture of how great Modern architects incorporate geometry into their work.

Kahn's stated views regarding proportion

A sense of Kahn's attitude towards proportion must be sifted from seemingly conflicting public statements. Regular *NNJ* readers may be frustrated by the present authors' decision to approach the topic of the Kimbell's proportions from what may seem to be quite a distance, by first looking at his broader theoretical concerns and at what can be thought of as Kahn's central theoretical statement, his article titled "Form and Design." It will hopefully become apparent how such an obtuse strategy will, in the end, provide a clearer picture of the building's proportions – or lack thereof.

According to Tim Vreeland from Kahn's office, "Form and Design" embodies Kahn's thinking better than any previous text [Vreeland 1961], and Kahn would not produce such a painstakingly considered text at any later time. David De Long claims that those inquiring about Kahn's theory would be routinely sent a copy of this article [De Long 1991, 71]. In it, Kahn argues that particular buildings of the same type share an archetypal essence, or "form," which is transcendent. He speaks of architects having mystical revelations of ideal "forms", then translating these into terrestrial buildings through a process he refers to as "design". According to Kahn's favourite illustration of his thesis,

... in the differentiation of a spoon from spoon, spoon characterizes a form having two inseparable parts, the handle and the bowl. A spoon implies a specific design made of silver or wood, big or little, shallow or deep. Form is "what." Design is "how" [Kahn 1961b, 145-154].

Kahn's terms "form" and "design" openly acknowledge the tension between the timeless aspects of architecture and the circumstantial processes that frame the design and construction of particular buildings.

Although it is undated, Kahn's earliest documentation of this thesis is most likely a hand-written draft within his personal notebook [Kahn 1959]. The first recorded public expression of this precise thesis is contained in a public address delivered at the Cooper Union entitled "The Scope of Architecture" on 20 January 1960 [Kahn 1960]. Kahn's preoccupation with the thesis dominated his theorising throughout 1960, leading to a Voice of America broadcast on 21 November 1960 [Kahn 1961a], the revised transcript of was be published in April 1961 as "Form and Design" in *Architectural Design* [Kahn 1961b] and reprinted in the book by Vincent Scully entitled *Louis I. Kahn* [Kahn 1962]. Kahn's distinction between universal "forms" and the circumstantial outcomes of the "design" process provides a basic framework by which to appreciate his statements regarding proportion.

Where proportion has traditionally been thought of as the terrestrial adumbration of an unseen realm, Kahn does not appear to associate proportions with the universal realm of "form". Rather, he associates proportions with the circumstantial "design" process. In the same year that he published "Form and Design", Kahn stated that, "[d]esign is a material thing. It makes dimensions. It makes sizes" [Kahn 1991a, 141]. "Form", on the other hand, "is not design, not a shape, not a dimension. It is not a material thing" [Kahn 1991a, 141]. By associating dimensions with the idiosyncratic "design" phase, Kahn appears to be saying that dimensions, and therefore proportions, are simply a matter of personal preference, and are subject to everyday concerns, such as cost or construction systems. If architects would like to establish mathematical relationships between key proportions, then they are free to, but in doing so they would not be adumbrating the universal realm of "form". This aspect of Kahn's theory leads the philosopher Arthur Danto to remark that his work is closer to "the spirit of Plato than architects whose buildings look like diagrams for geometric theorems" [Danto 1999, 188]. According to Danto, Kahn, like Plato, is more concerned with the essential elements required of such things as political states, beds and art museums, in order for these things to exist at all.

While, to Kahn's way of thinking, proportions may not point to anything higher than an architect's own taste, this does not mean that they could not form a vital aspect of Kahn's own sensibility. Unfortunately, Kahn's theoretical pronouncements provide no clear sense of his personal preference. On one occasion, Kahn specifically states his preference for buildings *without* mathematical proportions. He states that

... to make a thing deliberately beautiful is a dastardly act; it is an act of mesmerism which beclouds the entire issue. I do not believe that beauty can be created overnight. It must start with the archaic first. The archaic begins like Paestum. Paestum is beautiful to me because it is less beautiful than the Parthenon. It is because from it the Parthenon came. Paestum is dumpy – it has unsure, scared proportions. But it is infinitely more beautiful to me because to me it represents the beginning of architecture. It is a time when the walls parted and the columns became and when music entered architecture. It was a beautiful time and we are still living on it [Kahn 1986, 91].

In the context of this quotation, the phrase "to make a thing deliberately beautiful" refers to the application of sophisticated proportional systems to architectural compositions, since what differentiates the temples at Paestum from the Parthenon are their "unsure, scared proportions."

While not rejecting the use of proportions outright, Kahn subordinates this device to a sense of the "archaic." The Temples at Paestum are championed for their chronological – and, in a sense, their ontological – proximity to architecture's mythical beginnings as a poetic discipline.

While the above statement highlights Kahn's fascination with the origins of building types and institutions generally, it does not give a complete picture of his ruminations on the topic of proportion. For a more complete appreciation of Kahn's attitude, a number of statements regarding music must be taken into consideration.

For Kahn, the principles guiding architectural production apply equally to musical composition. This is made clear in a statement he makes in 1964, in which the relationship which he purports to exist between particular "designs" and their corresponding "forms" is extended to particular pieces of music and their underlying structures. "If I were a musician," Kahn argues,

> … and I were the first person to invent the waltz,
> the waltz doesn't belong to me at all,
> because anyone can write a waltz –
> once I say there is a nature of musical environment
> which is based on three-four time.
> Does that mean I own the waltz?
> I don't own the waltz
> any more than the man who found oxygen owns oxygen [Kahn 1991b, 177].

Kahn's feigned modesty in this quotation belies the fact that he was actually a talented musician. In his youth, and without the benefit of formal training, Kahn had helped support his family by playing the piano at a local cinema [Tyng 1984]. There is also evidence to suggest that he may have taken an interest in the traditional relationship between architecture and music. In February 1956, during a period of profound transition in Kahn's work, the architectural theorist Colin Rowe wrote to Kahn to thank him for an evening of intense discussion which they had spent together [Rowe 1956] and to inform Kahn that he would be sending him a copy of Rudolf Wittkower's book, *Architectural Principles in the Age of Humanism* (the 1952 second edition). A major portion of Wittkower's book discusses the musical consonants that regulate the proportions of Palladio's villas, a fact that Rowe may have considered when informing Kahn that in Wittkower's book, "I think that you will discover attitudes with which you are profoundly in sympathy" [Rowe 1956].

Keeping in mind that Kahn's theoretical pronouncements are often oblique (as though he were avoiding any binding contracts with himself), two of his statements do seem to echo Wittkower's analysis of Palladio. In the following statement, Kahn imagines a space having the character of a sound:

> I imagine myself composing a space lofty, vaulted, or under a dome, attributing to it a sound character alternating with the tones of the space, narrow and high, with graduating silver, light to darkness [Kahn, cited in Robinson 1997, 12].

In "Form and Design", Kahn also writes that "[t]o the musician a sheet of music is seeing from what he hears. A plan of a building should read like a harmony of spaces in light [Kahn 1961b, 149].

Kahn's conception of space in aural terms recalls the connection made by architects of the Neoplatonic tradition between spatial proportions and musical consonants, as described in Wittkower's book.[1] Indicative of one who boasts that he only tends to read the first few pages of books, Kahn appears to appropriate a simplified version of Wittkower's thesis into his own theory,

---

by attributing a sound character to space. Notably, in the introductory pages to *Architectural Principles*, Wittkower discusses Alberti's belief that "in music the very same harmonies are audible which inform the geometry of the building" [Wittkower 1971, 9].

At this point and despite Kahn's references to music, it is important not to overstate his interest in proportional systems. Kahn's "Form and Design" thesis confers a higher status to the inseparable combination of fundamental elements that constitute a type, while relegating dimensions to the so-called "design" process. In this sense, the act of inscribing proportions sits on par with relatively mundane tasks, like meeting a client's budget, or adhering to building regulations.

Yet there is another statement of Kahn's that leaves open the possibility that proportions might perform a higher, or numinous, role in his architecture. In "Form and Design," Kahn argues that a great building "must begin with the unmeasurable, must go through measurable means when it is being designed and in the end must be unmeasurable" [Kahn 1961b, 149]. One paragraph later, he reiterates this view, arguing that

> ... a building has to start in the unmeasurable aura and go through the measurable to be accomplished. It is the only way you can build, the only way you can get it into being is through the measurable. You must follow the laws but in the end when the building becomes part of living it evokes unmeasurable qualities. The design involving quantities of brick, method of construction, engineering is over and the spirit of its existence takes over [Kahn 1961b, 149].

Precisely how a building can evoke "unmeasurable" qualities is never spelled out. Other scholars have argued that Kahn does this through his control of light.[2] Yet there is no reason to exclude mathematical proportions from Kahn's repertoire of numinous devices.

From Kahn's complex, oblique and at times contradictory pronouncements, one plausible interpretation emerges, which – as will be seen shortly – seems to reckon with the mathematics of the Kimbell Art Museum. Firstly, it must be recognised that Kahn is more interested in fundamental spatial arrangements, and so the inscription of proportions is not so much an imperative as it is a matter of personal preference. Also, as something that would occur during Kahn's "design" phase, the process of making dimensions would at times be overridden by the client's budget, or other circumstantial factors, rather than the desire to achieve harmonious proportions. Having said this, there is evidence to suggest that Kahn might welcome the presence of proportions, especially those with a musical basis (such as 1:2, 2:3 or 3:4) – where they can be achieved in conference with competing circumstantial factors. Their presence could even evoke what Kahn calls "unmeasurable" qualities. Finally, in light of Kahn's remarks about the Parthenon, we can assume that he would be more interested in relatively simple ratios, and would be unlikely to delight in complex or overly sophisticated proportions. Notably, none of Kahn's statements champion proportions produced using geometry, such as $\sqrt{2}$ or $\sqrt{3}$, which, if they are to be found in his buildings, would need to be treated as either secrets, or accidents.

The 2' square grid

Part 2 of this paper will present the findings of a geometrical analysis of the Kimbell, highlighting Kahn's apparent use of approximate (though *very* close), $\sqrt{2}$ and $\sqrt{3}$ rectangles and their attendant grids in the overall planning of the Kimbell. However, in the context of the preceding textural analysis – which highlighted the importance to Kahn of music – a small number of precise harmonic proportions warrant closer investigation.

Anyone who has tried to design a building with an array of geometrically constructed proportions, would be aware of the tendency of such practice to produce very fussy dimensions. Indeed, any geometrically derived system will need to be compromised at the level of internal walls or fenestration, unless, of course, the architect is willing to calibrate their dimensions to the nearest half inch, or centimetre.

However, the dimensions of the Kimbell are anything but fussy. In plan, for example, they are exclusively in whole feet. The sanctity of this rule effectively precludes the existence of many *precise* ratios incorporating irrational numbers. Rather, the building contains a number of numeric, or modular, ratios involving small rational numbers. In plan, the internal proportions of the bays are 5:1. In section, the internal proportions of each bay are 5:3 (measuring to the springing point of the cycloid shells), while the internal proportions of the interstitial spaces linking each vaulted bay (or "servant" spaces as Kahn called them), are, in section, 1:2.

Such elementary proportions are made possible through a relentless adherence to an imperial module, the foot. Moreover, the plan conforms to a square grid, 2' x 2'. Were such a grid laid over the plan, it would pick up every column and wall. Even the floor pattern, comprising 6' x 2' travertine floor tiles, snaps to the grid. Smaller elements, such as 1' x 1' pieces of parquetry, or the 1" x 1" mosaic tiles in the toilet cubicles, can also be seen in the light of an all pervasive imperial module, in the service of simple drawings.

Further to his use of an imperial module, Kahn further simplifies dimensions by tending to use values with common factors. The 5:1 planar proportion of the vaults reflects the simplest of measurements – 100' x 20'. The sectional proportions of 5:3 reflect simple dimensions again, 20' x 12', while the 12' high "servant" spaces are simply 6' wide, yielding an harmonic proportion of 1:2. Similar combinations of whole feet dimensions, featuring numerical values with many factors, are to be found throughout the Kimbell.

Kahn's adherence to a 2' x 2' square module, the resultant commensurability of dimensions in the Kimbell, and the corresponding lack of *precise* proportions resulting from geometrical constructions, makes perfect sense when seen in relation to Wittkower's thesis regarding the architecture of the Italian Renaissance. In *Architectural Principles*, Wittkower states that the central issue of Renaissance architecture is the commensurability of ratios and that recent scholars obscure this fact "by insisting on the theoretical and practical advocacy of incommensurable, i.e., geometrical proportions by Renaissance architects" [Wittkower 1971, 158]. "It seems almost self evident," he argues,

> ... that irrational proportions would have confronted Renaissance artists with a perplexing dilemma, for the Renaissance attitude toward proportion [...] was aimed at demonstrating that everything was related to everything by number [Wittkower 1971, 158].

Since Kahn owned a copy of Wittkower's book, and since Wittkower's thesis is almost as valid for the Kimbell as it is for a Palladian villa, it would be reasonable to conclude that Wittkower's book had some effect on the Kimbell's dimensions. The fact that Renaissance architects believed harmonic proportions adumbrated a Platonic realm, in much the same way as Kahn thought buildings could evoke the "unmeasurable", adds further weight to this interpretation.

Archival evidence

However, one major obstacle stands in the way of this analysis, and that is the office correspondence for the Kimbell, which contains compelling evidence that the client, and not the architect, made crucial decisions regarding not only rough sizes or areas, but actual dimensions.

For a comprehensive account of the Kimbell's design and construction, based on archival records and interviews, the reader is referred to Patricia Cummings Loud's text, the *Art Museums of Louis I. Kahn* [Loud 1989]. Loud outlines an iterative and protracted process that typifies Kahn's working method. The present paper will only refer to a small number of key documents which relate specifically to dimensions.

The office correspondence reveals no record of a conscious attempt to establish harmonic consonants. On the contrary, it shows that many dimensions were established in conference with the client, the gallery director Richard Brown, who regarded himself as Kahn's "sparring partner" [Brown 1967b] and who persistently pressured Kahn to reduce the overall size of the gallery. His disposition can be summarised in a letter he wrote to Kahn on 12 July 1967 regarding a design iteration that was to be 400′ square:

> Within that big square you wind up with an awful lot of cubic <u>space</u> that must be heated, air-conditioned, illuminated, etc.; and acres of floor and wall surface that must be cleaned, waxed, mopped, resurfaced upon occasion, etc.; all of which costs money and labor to do, and I want as much money as possible saved from maintenance so I can buy more and more art as the years roll by, not just keeping up the house [Brown 1967a].

In the same letter, Brown raised what he calls the issue of scale, writing that "The Grand Canyon is vast and its scale is exactly right [...]; size and scale are in balance" [Brown 1967a]. He contrasts it with a small Rococo church, that helps the user "feel as secure and intimate with God and the universe as does a warm bath behind a door bolted against any possible intrusion" [Brown 1967a]. The letter goes on to state that Mr. and Mrs. Kimbells' paintings "are very 'gentile', 'polite' representations of fair ladies, tender little children and singularly pure young men" [Brown 1967a]. These arguments form a lengthy preamble to a list of size-paring instructions, including a suggestion that the gallery walls be lowered from 15′ to 12′, and that the upper most height of the vaults be lowered from 30′. Subsequent iterations of the design feature 12′ high walls as requested, and a much shallower vault, based on a cycloid. Astute readers will have noted that 12′ is the short side of the aforementioned 5:3 sectional ratio.

In a hand-written letter and sketch, dated 11 May 1968, Brown noted that the design at that date was as long as the Dallas International Airport terminal, a building "<u>notable for its affect of huge scale!!!</u>" [Brown 1968]. Effacing himself and his aversion to monumentality, Brown signed the letter "Richard the Chicken Hearted." In subsequent iterations, the gallery bays are reduced from 120′ to 100′. (At the simple level of building technology, it can also be observed that 100′ happened to have been the maximum distance that concrete walls or vaults could be produced without requiring expansion control joints). Under similar pressure from his client, Kahn reduced the width of the bays from 30′ to 20′. There is no evidence that Kahn chose this dimension for its 5:3 relationship to the 12′ wall height, or its 1:5 – that is, its modular – relationship with the length of the bays, though the fact that 20 is a factor of 100, and has the number 4 as a factor in common with 12, may have influenced his choice. This distance also reflects Kahn's belief – based on an assertion by the American architect Clarence Stein (1882–1975) – that 20′ is a minimum width for any space containing artworks. According to his wife, Kahn had not allowed artworks to be hung in their home because it was not 20' wide.

While there is no direct archival evidence to suggest that Kahn consciously sought harmonic proportions, the office files do testify to his commitment to the imperial module. The day-to-day negotiations and directions between Kahn, other members of the design team, and Richard Brown, were conducted and documented via written correspondence, and it would have been a matter of convenience that sizes be expressed in terms of whole feet. Unlike other kinds of communication, including the transfer of electronic drawing files, or the exchange of plans, letters are a poor means of communicating fussy dimensions. To reduce the likelihood of error, Kahn fostered a kind of written discourse in which dimensions were exclusively expressed in whole feet, and where, by implication, plans and sections would adhere to an invisible imperial grid.

A few examples of this rule have been seen in Brown's letters to Kahn, but other examples abound. A raft of major dimensional changes were negotiated in various letters between Kahn's associate Marshall Meyers and the partnering architects in Texas [Meyers 1969; Meyers 1967]. Smaller dimensions are discussed in another letter in which Meyers is asked to nominate which columns need to be 3' by 2' for structural reasons, as opposed to the 2' square column used generally in the project [Harden 1969]. Columns measuring 2'-6" are not considered, though they are likely to have been adequate structurally. In another letter, the partnering architects mentioned that the suspended floor slab could be 10" thick, with sheer heads at the top of each column [Gerin 1969]. Wishing to maintain the imperial module in section (and wishing to keep cleaner lines), Kahn opted for a thicker slab, 1' in depth.

Conclusions based on the textural analysis

Despite the burdens of modern practice, Kahn was able to accommodate traditional design parameters in his own work. He achieved harmonic proportions by limiting the way in which dimensions were communicated on a day-to-day basis in written correspondence. For architects, exchanging letters with clients and a broader design team is an everyday occurrence, whereas plans are exchanged far less frequently. In the case of the Kimbell Art Museum, one limitation of letters – that they cannot easily communicate fractional dimensions – contributed to that building's geometrical strength, by giving it a 2' square grid, which locks in an array of harmonic proportions.

Were it not for his client's fiscal constraints, Kahn would surely have made the Kimbell a much larger gallery. It might have had bolder proportions, such as a square section, but circumstances led to less striking proportions, based on musical ratios. The fact that any proportions resulted from the complex interplay of client's, consultants' and architect's input, can, at least in part, be attributed to Kahn's subtle manipulation of day-to-day correspondence. Where many commercial architects would sacrifice timelessness for timeliness, Kahn appears to have managed both.

When seen in terms of his theoretical pronouncements, it could be argued that Kahn strove for harmonic proportions as part of a quest to evoke an "unmeasurable" realm of inspiration. This is the realm of Kahn's so-called "forms", and can be likened to Plato's realm of ideal Forms. While nothing can be taken from Kahn's words to suggest that he was interested in geometrically-derived proportional systems, the fact that he engages with the Neoplatonic tradition on one front suggests that he may have engaged with other Neoplatonic concerns. This line of thinking will be pursued in Part 2 of this paper, which identifies the Kimbell's geometrically constructed proportions.

PART 2

Geometrical proportions in the Kimbell

Those who are familiar with recent publications about Kahn will be aware of Klaus-Peter Gast's book, *Louis I. Kahn: the Idea of Order* [Gast 1998]. In it, Gast argues that Kahn continues the Neoplatonic tradition by consciously inscribing his buildings with a hidden geometry, and by using the square and its derivations as a motif in his work. However, as Fleming has argued previously, Gast's analysis of the Kimbell is by no means exhaustive, and may contain inaccuracies [Gast 1998; Fleming 1999]. According to Gast, the distance by which the external dimensions of the rectangular plan of the Kimbell Art Museum falls short of being a double square, determines the width of the gallery's many bays. This claim simply does not tally with Kahn's working drawings, according to which the Kimbell is 318' wide (measuring from north to south) and 174' deep (measuring from east to west).[3] To be a double square, the building would need to be 348' wide, that is, twice as wide as its depth of 174'. The difference between its actual width and the width it would be were it a double square is 30' and this is the distance Gast refers to in his analysis as *x*, which should also be the width of the gallery's bays. However, the bays are not 30' wide. They are 20' wide – or 22' when measuring from the centres of the supporting columns. This represents a discrepancy of between 8' and 10' (or 40% to 50% of 20').

For the purpose of this study, it was deemed necessary to conduct a fresh geometrical analysis, and to inform the reader of its parameters and limitations. Padovan [1999] argues that the two major approaches in securing precise information regarding proportioning systems are numerical calculation and geometric construction or deconstruction. Whenever possible, both procedures should be utilized, and ideally, the analyst would also prefer *in situ* measurements personally carried out. However, with the advent of modern architectural drawing techniques, we do have fairly truthful drawings and measures carefully considered by the architect and associates, and we can reasonably consider the accuracy of the building's proportioning as written and drawn.

Additionally, there is general agreement that percentage deviations may be indicated in scholarly works to assist everyone in knowing how much the actual structure varies from the ideal proportioning system. The formula for determining percentage deviation is:

Measured (or Actual) – True (or Ideal)/True x 100

The deviation can be either high or low (+ or –). The understanding is that the closer the actual answer is to the ideal, the more the probability that the ideal measurement being considered was actually used. It isn't "proof positive", but it is suggestive of intention. The only time we are positive is when we hear or read about that intention directly from the creator of the work.

Using a procedure that Reynolds has previously shared with *NNJ* readers [2001], the overall plan was analysed, with specific attention to the internal volumes. In other words, arcs and compass points were focussed on internal wall faces, rather than column centres or external surfaces. The celebrated architectural space of the museum is framed by the internal surfaces of its walls, and so the thickness of those walls is of less significance.

The analysis began with the most elementary of analytical procedures, which identified, almost instantly, geometric proportions in the Kimbell's three major wings, those being the southern and northern wings each comprising five vaulted bays (excluding the porches), and the central wing, containing three vaulted bays (also excluding the porches). The five-bay northern and southern wings form rectangles of approximately √2 proportions. They have internal dimensions of 100' x

140', yielding a ratio of 1:1.4. This falls very close to a √2 rectangle, of 1:1.414. The three-bay central wing has internal dimensions of 86' x 100', yielding a ratio that falls very close to the rectangle of the equilateral triangle, which is .866025:1. The fact that the eastern-most vaults of the northern and central wings are partitioned off from the gallery, making these ratios imperceptible to the viewer, raises the possibility that the presence of these ratios is a matter of coincidence. Were these ratios precise to a number of decimal points, then it would also be easier to argue that they are intentional. This is, after all, a building made using reinforced concrete technology, which could easily accommodate precise dimensions. However, achieving precise dimensions would also have required very fussy dimensions, and, as we saw in Part 1 of this paper, the Kimbell's dimensions are anything but fussy. They are exclusively in even numbers of whole feet. If we accept the pre-eminence and practical benefits of a 2' grid, then the percent deviations between the actual and ideal proportions mentioned are less significant than they may at first appear. For instance, each is accurate to the nearest foot, if not a fraction thereof.

The next step in the analysis was to pursue a simple hypothesis; in deriving the final form of the southern wing, Kahn may have exploited one of the primary features of a √2 rectangle, that, when it is divided in half on its long side, each half is also a √2 rectangle. The √2 rectangle is the only rectangle that does this.[4] The following analysis of the southern wing does indeed support that hypothesis, revealing what some readers will agree is a reasonably elegant marriage between a classical proportioning system and the final built form.

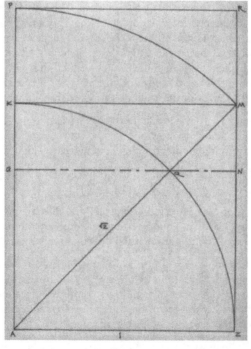

Fig. 1

The √2 rectangle is made from the side and the diagonal of the square. As seen in fig. 1, when the side of the square AZ is 1, the diagonal AM is equal to √2, or 1.414... . We then generate the

√2 rectangle APRZ. This unique relation between side and diagonal in the square is further developed when we find that where the side of the square passes through the diagonal, at Q, this interaction divides the rectangle exactly in half at GN, precisely into two √2 rectangles, AGNZ and GPRN. Looking solely at the square, this intersection would yield a reciprocal √2 rectangle, 1/√2 or 0.707… /1, within the square. This division at Q yielding the line GN was referred to as the "Sacred Cut" by Tons Brunés [1967].

In this same construction, it is also essential to note that rectangle KPRM that sits over the square AKMZ is the "theta", or θ, rectangle. Its ratio is 1:2.414…, or a √2 with a square added to the *short* side of the √2 rectangle. It is interesting that the θ rectangle contains the two component parts of the generated √2 rectangle.

In fig. 2, the technique of "Rebattment" is shown. Rebattment is taking the short side of a rectangle and rotating it onto the long side in order to "cut off" a square in that rectangle. Here, AZ is rotated onto AP and ZR to produce the square, AKMZ, and PR is placed onto those same two long sides to produce the square GPRN. In this figure, the pairs of diagonals, AM with ZK and GR with NP, have been drawn in to clarify these two squares, whose centers are O and Q. When this rebattment technique is done in the √2 rectangle, the squares overlap to produce the specific rectangle GKMN. This is the 1.707… This rectangle, like the θ, is also a square and a √2, but here, the square has been applied to the *long* side of the √2, not the short side. As a result of the 1.707 in the middle, θ rectangles are generated in the top (KPRM) and the bottom (AGNZ).

Fig. 2

When diagonals are generated in the two θ rectangles, we locate their centers, X and Y, as can be seen in fig. 3. When midlines EF and CJ are drawn through these centers, they generate the square ECJF. It is generated into the center of the √2 rectangle APRZ. The center O is the center of the square *and* the √2 rectangle. It is critical to note that these qualities are unique to the √2 rectangle, that they are key elements in the √2 proportioning mechanism. These particular elements are:

–   The square, 1 x 1
–   The √2 rectangle, 1.414… x 1
–   The θ rectangle, 2.414… x 1
–   The square plus the 1/√2, 1.707… x 1

Fig. 3

In any rectangle, by developing the basic elements in squares and rectangles, various grids and intersections are created. It is these intersections and interior line lengths, called *caesurae* (as in music, they are breaks) that develop the proportioning systems that are unique to the specific ratio being utilized. In the √2 rectangle, the square, √2, θ, and the 1/√2 are the primary elements found in the grid system. These intersections and caesurae are then used to design and develop the space, the specific compositions and arrangements that we see. Each architect, designer, and artist is free to develop particular grids for specific needs. Should the Kimbell have been geometrically regulated, then what choices might Kahn have made for his √2 (and, as we shall see, the √3) rectangles?

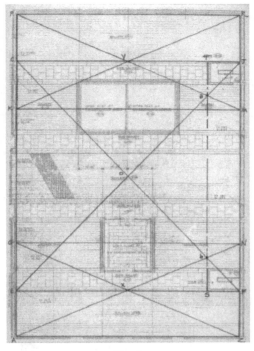

Fig. 4

A plausible utilization of the aforementioned grid systems in laying out the upper level plan of the Kimbell, is described in fig. 4. The centrally-placed master square ECJF of the grid provides for the placement of the two side galleries, as can be seen along the top CJ and the bottom, along GN, of the square. Furthermore, the intersections of the diagonals FC and EJ of this square, with the diagonals of the two θs, along AN and MP, generate the "eyes", or intersections, at points **a** and **b**. The caesura ST is drawn through these points, and define the ends of the two staircases with the edges SF and TJ. (We will see the completion of these stairwells in the next grid.)

The development of the entire southern wing may have followed a similar procedure. In fig. 5 it can be seen that:

- The rebattment technique may have been used to create the two master squares, AKMZ and GPRN.

- The two θ rectangles AGNZ and KPRM could have been defined with their diagonals in order to generate the third master square ECJF in the center of the rectangle.

- The diagonals of all three squares have been drawn in.

- The grid is now complete.

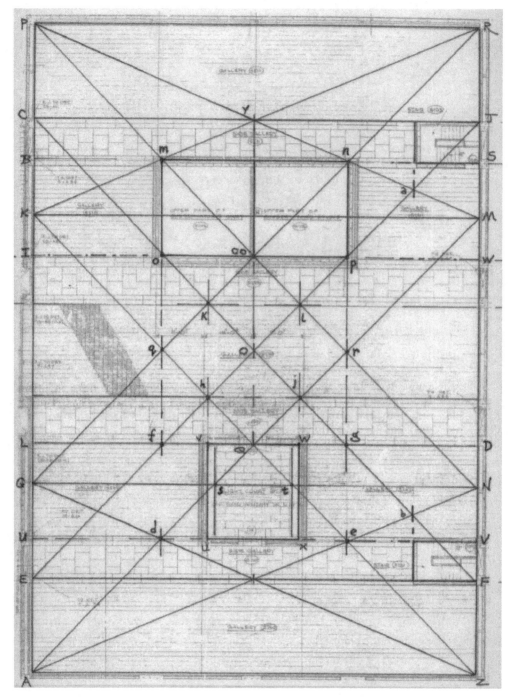

Fig. 5

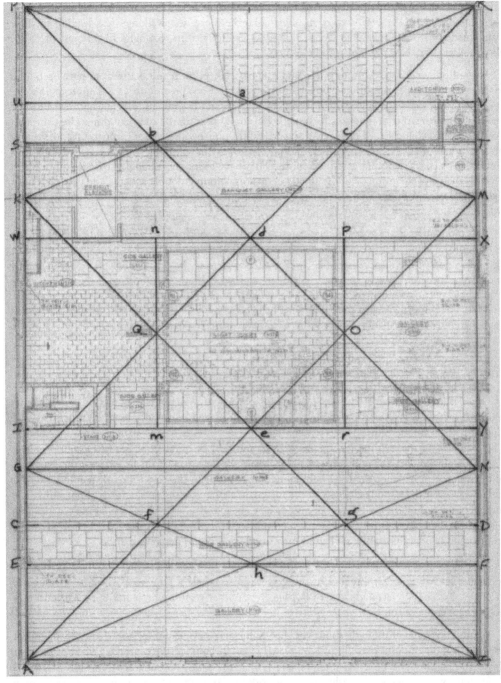

Fig. 6

While on the topic of the stairs, it is worth noting that their rectangular area appears to have been completed by the points **d, e, m,** and **n**. Furthermore, the walking areas leading to these stairs, made of stone rather than wood, along the very long, thin rectangles EUVF and BCJS are also developed by these particular grid elements.

Also, of points **d, f, q, o, m, n, p, r, g** and **e**, specifically **o, m, n** and **p** appear to have been used for the double square **omnp** the upper level of the conservator's courts. All these points are generated from the diagonals of the three master squares and the two θs, top and bottom. The line **op** that completes this double square is made by simply passing the caesurae **dm** and **en** through the diagonals, FC and EJ, of the central square, ECJF. Line **op** also passes through OO, the center of the upper master square, GPRN. Kahn continues his unifying thought by making this conservator's court two squares. These courts could have been any rectangle he desired, but Kahn kept his theme of the primacy of the square by making **omnp** a double square. The point has been made by Gast and others that squares are both a motif and generative figure in much of Kahn's work.

This thought is an apt entrée to an analysis of the light court on the western side, center (at the bottom of the wing). This court is also a square **uvwx**. This space appears to have been made from an intersection of the diagonals of the three master squares of the √2 rectangle. The points **h, k, l** and **j** were used for the left and right edges, **uv** and **xw**. The westernmost edge **ux** is defined by UV. The easternmost edge **vw** is made from the horizontal midline LD of the base master square AKMZ. The center Q is also the midpoint of the edge **vw**. The points **s** and **t** further define this court's architecture elements. Points **s** and **t** are made from the intersection of the base GN of the top master square GPRN.

Furthermore, these court openings could be linked by being at the opposite vertexes Q and OO of the dynamic square QqOOr. The internal square **hklj** is in fact the same size as the light court.

Fig. 6 examines the northern bay. It is also an approximate √2, but is very slightly larger than its southern mate. Also, a slightly different grid appears to have been used. Here, the two θs AGNZ and KPRM frame the central area GKMN, which is the 1/√2 we have already mentioned above. The diagonals of the two master squares AKMZ and GPRN intersect at points Q and O. These points, combined with the points **b, c, f** and **g**, where the diagonals of the θs intersect the diagonals of the squares, provide for the large light court square **mnpr** in the very center of the floor plan. It can also be seen the various changes in the materials of the walking spaces of the floor utilize the intersections within the two θs, at points, **f, h, g, b, a** and **c**, as well as the centers of the two squares **e** and **d**. The width EG is but one example. In both √2 bays, the grids are relatively simple and elegant. More importantly perhaps is how simple they are to perceive.

Another way of looking at the grid in this North bay is to see WPRX as a double square (√4), rectangle EWXF as a √2, and to see the very wide rectangle at the base AEFZ as a double θ (1:4.828...). The √2 rectangle is very flexible; yet another reason why it has been a favourite of designers over the centuries.

Finally, my original thought about whether Kahn may have exploited the characteristic aspect of √2, that it divides into two √2s, was answered when I examined the central bay. It measures out to be 100' x 86'. The point of interest here is that the ratio, 0.866... to 1 (or 1.154... to 1, depending on whether you're using the short side or long side as your unity; these are simply reciprocals of each other) is known as the "rectangle of the equilateral triangle". It is also what is known as *ad triangulum* geometry. In what may betray a sense of humour geometrically, Kahn appears to have selected the rectangle that consists of *two* √3s and interjected it between two √2s! If he did, then he achieved a clever, elegant, and ingenious solution to a plan that calls for variation within symmetry, that the central rectangle be in harmony with, but different than, its flanking bays on the northern and southern sides.

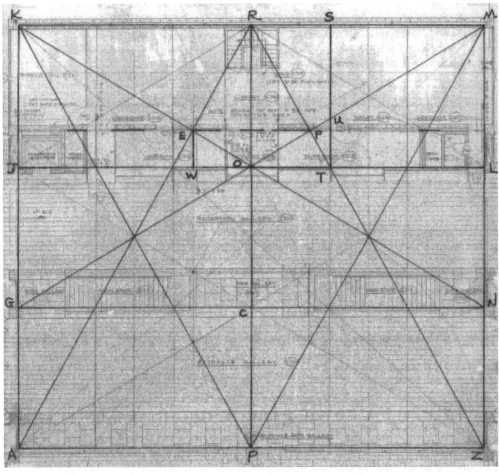

Fig. 7

The two √3 rectangles are tangent on their *long* sides. In Figure 7, the rectangle AKMZ is the *ad triangulum* rectangle, with rectangles AKRP and PRMZ both being √3s. ARZ is the equilateral triangle, and so is KPM. By bisecting (again, the theme of two) the angles A, Z, K and M, two other √3 rectangles AJLZ and GKMN are also generated. The underlying materials on the floor are placed along this √3 grid, and are also still yet aligned with their √2s on either side. Of note is that the √3 contains three √3s within its perimeter, just as the √2 has two √2s. The agreement of these interior rectangles with the number under the radical sign is a quality of all root rectangles (a √4 has four √4s within, and so on). Looking to the top right √3 rectangle ORML a smaller √3 rectangle ORST can be seen. Point F is where a diagonal TR and a reciprocal line OU intersect. These intersections are sometimes referred to as "occult centers" because they are "hidden from the eye". In this grid, they provide for the width EW, as point E is also an occult center. As can be seen, EW becomes an integral part of the grid. And in what is perhaps most interesting, in a √3 rectangle the occult centers break the rectangle into *fourths*. In a √2, the occult centers divide the rectangle into *thirds*. This is another quality of the root rectangles: whatever the number is under the radial sign, the occult centers always yield the next number up fractionally. So, EW is one fourth of OR. Fourths bring us back to squares: 2^2, or 2 x 2, is four. Four happens to be the number of root rectangles used in the plan: two √2s and two √3s. Like many artists, designers, and architects at the time, Kahn may have known all these things though popular books such as the widely circulated the *Elements of Dynamic Symmetry* [Hambidge 1967].

Given the many unknowns, including the range of possible applications of √2 and √3 grids, this analysis cannot be considered definitive. Neither can its weakness be disguised. Some readers might argue that an experienced geometer can find proportions where they don't exist. It remains problematic also that the easternmost vaults of the northern and central wings are partitioned off from the gallery, rendering proportional relationships in these wings imperceptible to the viewer.

Within the confines of a research paper, it would have been impossible, and tedious, to have presented the hundreds of accompanying percentage deviation calculations that would complete this analysis; and indeed, it needs to be said that – with the exception of the squares, which are precise – the proportional relationships that have been identified here are not true to more than one decimal place. Close inspection of the figures reveals that some construction lines miss their relevant wall surfaces by up to (but at least no more than), one foot. In other words, the grid fits as well as could be expected given the overriding presence of a 2′ square grid and a whole foot dimensioning system; and that really is the salient point. The extremely close alliance of the central wing to a √3 rectangle, which is then flanked by near perfect √2 rectangles, and the approximate conformity of the internal layout to grids which proceed naturally from those figures, presents what amounts to an object lesson in proportioning when tempered by a dimensional grid. It also presents a scholarly conundrum. If the presence of approximate √2 and √3 proportions, and their attendant grids, is the result of what must have been a tireless effort by Kahn or members of his staff, then no record, either oral or written, has been found to corroborate such a thesis. If this analysis is correct, then why didn't Kahn mention his fascination with classical proportioning grids? Perhaps one of his office staff added this layer of refinement? Amidst the many tasks facing the design team leading to the production of construction drawings, could the act of refining a plan have taken so little time as to not warrant mentioning, thus leaving no written trace of the event?

Conclusions based on the geometrical analysis

With the Kimbell Art Museum, Kahn managed to reconcile a number of competing demands. First and foremost, this is a building in which technological ambition meets head-to-head with fiscal constraint. The building's roof is made from post-tensioned curved concrete beams spanning an incredible 100′. Within that complex roof structure, a sophisticated network of ducts and electrical services is carefully integrated, and remains virtually imperceptible to the viewer. The building meets a number of other demanding criteria related to security, egress, access, catering, archiving, the preservation of precious artworks, and, most significantly, a limited budget.

Yet Kahn, or whoever might have been responsible for incorporating *ad quadratum* and *ad triangulum* systems, appears to have taken the time to, in effect, salute two of the most famous root rectangle systems used in antiquity and the classical periods in art history. This would seem to be a totally appropriate geometric context in which to provide for the geometric structure of an art museum.

Notes

1. Inspired by figures including Pythagoras, Plato and Saint Augustine, architects of the Neoplatonic tradition applied to buildings the same proportions that underpin musical harmony, in the belief that those proportions are of transcendent origins and order the cosmos.
2. At least two scholars state that Kahn's control of sunlight has a numinous, or transcendent effect, See [Geldin 1991, 17]; see also [Dean 1978, 87].
3. See the drawing titled: "A4: Upper Level Floor Plan", Louis I. Kahn Collection, University of Pennsylvania and Pennsylvania Historical and Museum Commission.
4. A wide variety of rectangles used throughout history, like the golden section, the √3, and the whole number/musical ratios, have particular qualities that set them apart from all other rectangles. These qualities are then used as "tools" in various proportioning systems and grid structures. LeCorbusier's golden section-based Modulor system is but one more recent example.

References

BROWN, Richard. 1967. Letter, Brown to Kahn, 12 July 1967, L.I.K. Box 37, "Dr. R. Brown – Correspondence 1.3.66 – 12.70." Louis I. Kahn Collection, University of Pennsylvania and Pennsylvania Historical and Museum Commission.

———. 1968a. Letter, Brown to Kahn, 11 May 1968, L.I.K. Box 37, "Dr. R. Brown – Correspondence 1.3.66 – 12.70," Louis I. Kahn Collection, University of Pennsylvania and Pennsylvania Historical and Museum Commission.

———. 1968b. Letter, Brown to Kahn, 29 July 1968, L.I.K. Box 37, "Dr. R. Brown – Correspondence 1.3.66 – 12.70," Louis I. Kahn Collection, University of Pennsylvania and Pennsylvania Historical and Museum Commission.

BRUNÉS, Tons. 1967. The *Secrets of Ancient Geometry and their Uses.* Copenhagen: Rhodos Publishing Co.

DANTO, Arthur C. 1999. "Louis Kahn as Archai-Tekt." Pp. 185-204 in Arthur C. Danto, *Philosophizing Art: Selected Essays.* Berkeley: University of California Press.

DEAN, Andrea O. 1978. A legacy of light. *AIA Journal* 67, 6: 80-89.

DE LONG, David. 1991. "The mind opens to realizations: Conceiving a new architecture, 1951-61. Pp. 50-77 in *Louis I. Kahn: In the Realm of Architecture,* eds. David Brownlee and David De Long. New York: Rizzoli.

FLEMING, Steven. 1999. Klaus-Peter Gast, Louis I. Kahn: the idea of order (Book review). *Architectural theory Review* 4, 2: 116-117.

GAST, Klaus-Peter. 1998. *Louis I. Kahn: the Idea of Order.* Trans. Michael Robinson. Berlin: Birkhäuser.

GERIN, Preston M. Letter, Preston M. Gerin to Marshall D. Meyers, 26 March, 1969, L.I.K. Box 37, "Preston M. Gerin – Correspondence 11.11.66 – 7.21.69." Louis I. Kahn Collection, University of Pennsylvania and Pennsylvania Historical and Museum Commission.

GELDIN, Sherri. 1999. Prologue, Louis I. Kahn: Compositions in a fundamental timbre. Pp. 15-17 in *Louis I. Kahn: In the Realm of Architecture,* eds. David Brownlee and David De Long. New York: Rizzoli.

HAMBIDGE, Jay. 1919. *The Elements of Dynamic Symmetry.* New Haven: Yale University Press. Rpt. New York: Dover Books, 1967.

HARDEN, T.E. 1969. Letter, T.E. Harden, Jr. To Marshall D. Meyers, 20 June, 1969, L.I.K. Box 37, "Preston M. Gerin – Correspondence 11.11.66 – 7.21.69." Louis I. Kahn Collection, University of Pennsylvania and Pennsylvania Historical and Museum Commission.

KAHN, Louis. ca. 1959 (undated). Louis I. Kahn notebook. Box K12.22. Louis I. Kahn Collection, University of Pennsylvania and Pennsylvania Historical and Museum Commission.

———. 1960. The Scope of Architecture at the Cooper Union Hall, 1-20-60. Cassette recording. Louis I. Kahn Collection, University of Pennsylvania and Pennsylvania Historical and Museum Commission.

———. 1961a. Structure and Form. *Voice of America Forum Lectures,* Architecture Series, no. 6. Washington D.C.: Voice of America.

————. 1961b. Form and Design. *Architectural Design* **31**, 4 (April 1961): 145-154.

————. 1962. Louis I. Kahn, "Form and Design." Pp. 114-121 in Vincent Scully, *Louis I. Kahn*, New York: Brazillier.

————. 1986. Wanting to be: the Philadelphia school. Pp. 89-92 in *What Will Be Has Always Been: the Words of Louis I. Kahn*, Richard Saul Wurman, ed. New York: Access Press Ltd. and Rizzoli.

————. 1991a. The nature of nature. Pp. 141-144 in *Louis I. Kahn: Writings, Lectures, Interviews*, Alessandra Latour, ed. New York: Rizzoli International Publications Inc.

————. 1991b. Louis I. Kahn, Talks with students. Pp. 154-190 in *Louis I. Kahn: Writings, Lectures, Interviews*, Alessandra Latour, ed. New York: Rizzoli International Publications Inc.

LOUD, Patricia Cummings. 1989. The *Art Museums of Louis I. Kahn*, Durhum: Duke University Press.

MEYERS, Marshall D. 1967. Letter, Marshall D. Meyers to Mr. Preston M. Gerin, 21 December 1967, L.I.K. Box 37, "Preston M. Gerin – Correspondence 11.11.66 – 7.21.69," Louis I. Kahn Collection, University of Pennsylvania and Pennsylvania Historical and Museum Commission.

————. 1969. Letter, Marshall D. Meyers to Mr. Thad Harden, 20 May 1969, L.I.K. Box 37, "Preston M. Gerin – Correspondence 11.11.66 – 7.21.69." Louis I. Kahn Collection, University of Pennsylvania and Pennsylvania Historical and Museum Commission.

PADOVAN, Richard. 1999. *Proportion: science, philosophy, architecture*, New York : E & FN Spon.

REYNOLDS, Mark. 2001. Introduction to the Art and Science of Geometric Analysis. *Nexus Network Journal* **3**, 1: 113-121.

ROBINSON, Duncan. 1997. The *Yale Centre for British Art: A Tribute to the Genius of Louis I. Kahn*. New Haven: Yale University Press.

ROWE, Colin. 1956. Letter, Colin Rowe to Louis I. Kahn, 7 February 1956, file labelled, "Correspondence from Universities and Colleges", L.I.K. Box 65. Louis I. Kahn Collection, University of Pennsylvania and Pennsylvania Historical and Museum Commission.

TYNG, Alexandra. 1984. *Beginnings: Louis I. Kahn's Philosophy of Architecture.* New York: Wiley Interscience Publications.

VREELAND, Tim. 1961. Letter, Vreeland to Pidgeon, 11 January 1961 "Master File, November 1 through December 30, 1960," Box L.I.K. 9, Louis I. Kahn Collection, University of Pennsylvania and Pennsylvania Historical and Museum Commission

WITTKOWER, Rudolf. 1971. *Architectural Principles in the Age of Humanism*. New York: W. W. Norton and Company.

About the authors

Steven Fleming lectures in the history and theory of architecture at the University of Newcastle, Australia. He is presently a visiting scholar in the Department of Philosophy at Columbia University. He received his Ph.D. in 2003 from the Department of Architecture at The University of Newcastle, with a thesis on Classical Platonism with respect to Louis I. Kahn's concept of "form". He has worked as a practicing architect in Australia and in Singapore.

Mark Reynolds is a visual artist who works primarily in drawing, printmaking, and mixed media. He received his Bachelor's and Master's Degrees in Art and Art Education from Towson University in Maryland. He teaches geometry for art and design students, and sacred geometry and geometric analysis for graduate students at the Academy of Art University in San Francisco, California. Mark is also a geometer, and his specialties in this field include doing geometric analyses of architecture, paintings, and design. He also lectures on his work in geometric analysis at international conferences on architecture and mathematics. Two of the most notable were at the Nexus Conferences in Architecture and Mathematics in Ferarra, Italy, in 2000 ("A New Geometric Analysis of the Pazzi Chapel in Florence"), and in Mexico City ("A New Geometric Analysis of the Teotihuacan Complex") in 2004. For more than twenty years, Mr. Reynolds has been at work on an extensive body of drawings, paintings, and prints that incorporate and explore the ancient science of sacred, or contemplative, geometry. His work is in corporate, public, and private collections throughout the United States and Europe. In 2004, Mr. Reynolds had a drawing selected for the collection of the Achenbach Foundation of Graphic Art in the California Legion of Honor, and had 43 drawings accepted into the permanent collection of the Leonardo da Vinci Museum and Library, the Biblioteca Communale Leonardiana, in Vinci, Italy. Mark's work can be seen at http://www.markareynolds.com.

DJP Marshall II

Indiana University – Purdue
University Fort Wayne
229 ET Bldg - IPFW Campus
Fort Wayne, IN 46805 USA
marshald@ipfw.edu

Origins of an Obsession

Though many geometric shapes can be constructed from circles, this paper is about the geometric square. It will be demonstrated that while the square is not the easiest of the polygons to construct initially, it is both easily enlarged and easily subdivided. Step-by-step manipulations of the square provide an explanation for the architectural design of the Forum of Augustus.

Introduction

My fascination with the use of geometry in architecture began in 1978 during a month-long visit to Rome as a University of Kentucky architecture student. It was then that I "discovered" the geometrically rich Pantheon while in hot pursuit of cheap postcards.

This chance visit was my first experience of the sublime visceral architecture that is the Pantheon and the beginning of an ongoing fascination with the use of geometry in architecture. When I recently found myself back from Rome again in the summer of 2001 with a group of students from Indiana University–Purdue University in Fort Wayne, I began studying this use of geometry in depth.

But much to my surprise there was little information available on the step-by-step use of geometry in construction and even less information available to indicate the architectural reasons for any particular geometric arrangement in a design. I nonetheless continued to read and experiment with the analysis of published drawings by means of compass and straightedge while working to maintain, as much as possible, the perspective of a first-century architect, that is, the use of compass and straightedge for design and string and stakes for building and site layout. This paper is a result of that investigation.

One aspect of the investigation was my conclusion that, given ancient technology, it is likely that the tools and techniques used to design on paper are similar in purpose and technique to those used in laying out a building on its site. The tools and techniques used to apply the geometry in both cases would work in essentially the same way.

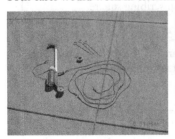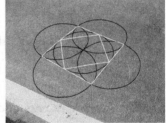

Fig. 1. One man, "stakes-string-chalk-tape", twenty-two minutes

Confident that valuable insight would be gained from my personal use of both sets of tools, I decided to explore the use of stakes, chalk, tape, and string to mark on the ground what I had drawn in my office using compass, pencil, and straightedge. Fig. 1 shows a few of the photos that were taken to record this experiment. The layout of this rotated square within a square seen in fig. 1 was completed single-handedly by the author in twenty-two minutes with considerable accuracy.

This experiment reinforced my belief in the similarity of techniques used for "office" design and site layout in earlier times.

The Square derived

Though many geometric shapes can be constructed from circles, this paper is about the geometric square. It will be demonstrated that while the square is not the easiest of the polygons to construct initially, it is both easily enlarged and easily subdivided. Its derivation from circles is as follows:

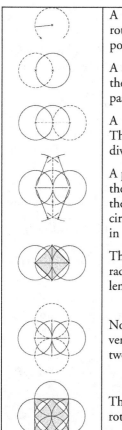

A first circle is created by holding a point at one end of a string and rotating a marker attached to the strings other end all around the fixed point.

A second circle is added which has its center point at the left quadrant of the first circle. It is rotated around this point with its circumference passing through the first circles center point.

A third circles center point in located on the first circles right quadrant. This creates a horizontal line through all three circles center points dividing the original circle in half.

A perpendicular line to divide the original circle into quarters is created by the standard geometric technique of bisecting a line. A line drawn from the bisector point to the center point of the original circle divides the circle into four equal parts and locates the four points of a square inscribed in the circle.

This first and smaller square can now be shaded. With the original circles radius of 1.0 and an area of 3.14, the square has an area of 2.0 and a side length of 1.414

Now set similarly sized circles having center points at each end of the vertical bisector. The intersections of the two new circles with the previous two circles are the four points of a larger circumscribing square.

The larger shaded square is exactly twice the area of the smaller square and rotated 45°. This is commonly referred to as a rotating square.

Subdivisions and extensions of the square

The extensive and repeated use of squares in the built environment, though economic because identical shapes are used for wall and corner construction, can rapidly become a mind-numbing grid without distinction, uniqueness, or attraction. In order to maintain visually interesting surroundings, harmonious manipulations of the square are necessary. When seeking these geometric manipulations, I felt strongly that an effort should be made to harness the creative possibilities found in small sets of interrelated rules such as the seven Newtonian colors of the painter's palette and the twelve harmonic tones that are the foundation of a Liszt concerto. It is my belief that just such a set of synchronized geometric techniques, which I will now describe, would have an inherent potential to create any number of designs, each of which could be imbued with a sense of resonant wholeness.

Having derived the square from a circle and now embraced it as the basis for further development, some ancient geometer probably began experimenting with string and stakes to determine what other shapes could be derived from the square. Purity would naturally be associated with those shapes most easily generated and so, therefore, preference was given to those shapes defined with as few as two strokes.

Following are the results of this explication of the square with notes indicating commonly used designations. In this paper, however, I will not be using these designations, which I feel would carry some baggage and unnecessarily encumber a fresh look at the subject. I will be using designations such as "1.414 technique" that clearly indicates proportions and avoids highlighting any of them as more "sacred" than the others.

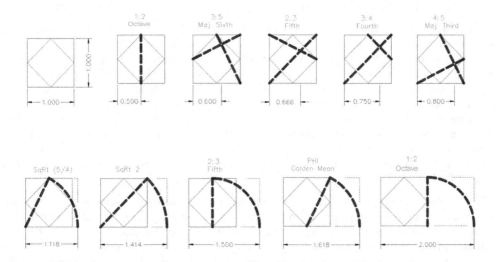

Fig. 2. The square's five subdivisions and five extensions

As can be seen from fig. 2, from the ten double-stroke derivatives of the square, many historically significant geometric ratios were anticipated. There are the 1:2, 2:3, 3:4, 3:5, and 4:5 small fraction ratios promulgated by Plato, Vitruvius, and Palladio for both music and architecture. Also shown are extensions such as the Sacred Cut ($1/\sqrt{2} = 0.707...$), and the Golden

Section ($\left(\dfrac{\sqrt{5}+1}{2} \right)$ = 1.618...). Each of the five subdivisions was performed with two lines, and

each of the extensions was accomplished by means of one line and one arc. Clearly a remarkable variety of shapes was possible for a pre-numerical geometer working single-handedly using strings and stakes.

The test

Now we are ready to test the hypothesis that beautiful, visually resonating structures can be built with no greater a set of geometric techniques than what are being proposed in this paper. Am I thus claiming that any structure built with these techniques will be beautiful? Not a chance! What I am proposing is that a reasonably creative designer would have a better chance of designing a compositionally cohesive and visual attractive structure using this small interrelated set of geometric techniques than if that same designer were left to their own "inspired" devices. Exceptions exist, of course, but probably fewer than are thought.

Precision and accuracy

When trying to "prove" a reconstruction, comparison and agreement become two of the primary tools by which a reconstruction gains credibility. But even a credible comparison is not certain proof. There is always a possibility that your explanation may be apparently correct while actually wrong, as was Ptolemy's explanation for the movement of the stars in his earth-centric explanation of the cosmos. However, predictive exactitude is still necessary even not sufficient.

In archaeological reconstruction the better the condition of the structure the more likely accurate dimensions can be established. Even so, as a practicing architect I know there is a considerable difference between an architect's original idea and the constructed building. It begins with a pure idea that the architect struggles to get on paper. That paper is given to a contractor who struggles to build it accurately. The building then lives a life of shifting foundations and adjustments. An archaeologist sometime discovers the remains of the structure and posits a reconstruction. The reconstruction is published at a greatly reduced scale and finally someone like me photocopies it for study. The opportunity for error is vast when attempting to represent the original idea of the building, even when the process is rigorously performed. For this reason I expect only reasonably approximate correlations between my reconstructions and the published images of others' proposals.

I have therefore adopted a process of collecting as many different representations of a given structure as possible and copying them to the same scale. Then using the tools and perspective of a first-century Roman architect, I study the building images using compass and straightedge. It is common in my experience for various reconstructions to have differing dimensions but also typical for them to share many aspects. It is at this point that I make a preliminary determination of the most likely shape and point of beginning for the reconstruction process. I then proceed, sometimes by chance and sometimes by plan, selecting those steps in the design that fit the criteria of being contiguous, continuous, and accurate.

This process has often begun with a one-unit square drawn using AutoCAD software and does so here. Using this software for drawing subdivisions and extensions gives me the capability of drawing with precision to eight decimal places. This readily allows the quick supporting or disproving of prospective steps in the process.

When overlaying existing published images with proposed step-by-step constructions, it is reassuring to know that my reconstruction is not exaggerated in a way as to support a predetermined outcome. The actual comparison process consists of proportionally enlarging the proposed reconstruction and overlaying it on an existing image for a visual analysis of its correspondence.

Forum Augustum

In order to rigorously test the preceding hypothesis it was decided to use a well documented and geometrically rich Roman imperial structure as a proof of concept. The Forum of Augustus was chosen as the example and a step-by-step examination of the design process was planned.

From the beginning of my study it was clear that Augustus's decision to locate his Forum alongside the Forum of Julius Caesar was not an accident. It is likely that the same political calculations[1] that placed a statue of Caesar inside Augustus's Temple of Mars, guided Augustus in placing his forum in close proximity to Forum Julium to maximize his own forum's "authority by association".

After much experimentation I decided that the beginning shape for generating Augustus's forum was a square located adjacent to Julius Caesars Forum and shown as dark dashed lines in fig. 3.

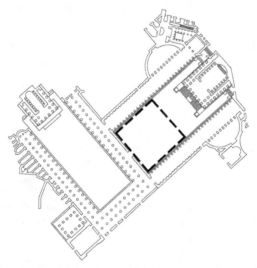

Fig. 3. Forums Julium (lower left) and Augustus (upper right)

One factor contributing to this decision was that the square shown is approximately 194 Roman feet, compared to the 200 Roman foot initial square of a military camp [Polybus 1979]. From this it seems plausible that Augustus, the General, might depend on familiar military layout techniques, or technicians, when designing his forum. Somewhat supportive of this decision is the comment on military camps by Josephus "within are streets and tents, those of the commanders being in the middle, and in the midst of all is the generals tent, in the nature of a temple" [Josephus 93 A.D].

The process

In describing the following process I am not attempting to establish the one and only process or ideal design of the Forum, but rather I am demonstrating just one of many harmonious designs possible with the set of techniques already noted. Many designs could have been adapted to this project's constraints during the step-by-step design process. But in the end, my decision to begin the design process adjacent to the Forum Julium was a more geometrically productive and politically significant point of beginning.

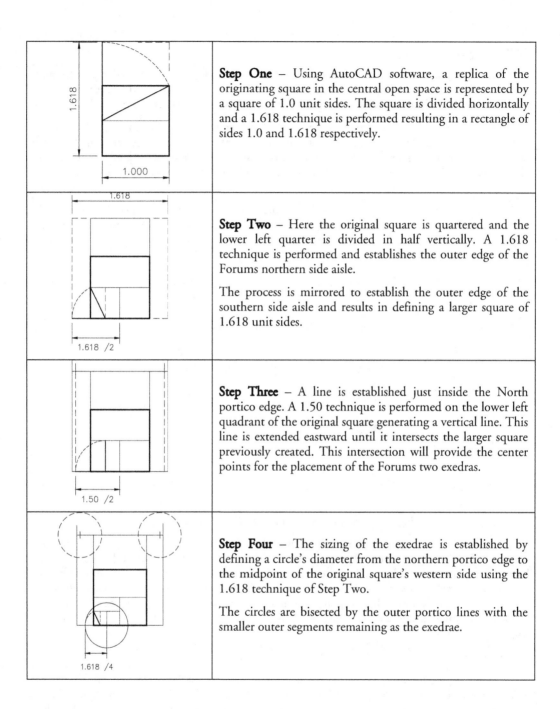

1.618 / 1.000	**Step One** – Using AutoCAD software, a replica of the originating square in the central open space is represented by a square of 1.0 unit sides. The square is divided horizontally and a 1.618 technique is performed resulting in a rectangle of sides 1.0 and 1.618 respectively.
1.618 / 1.618 /2	**Step Two** – Here the original square is quartered and the lower left quarter is divided in half vertically. A 1.618 technique is performed and establishes the outer edge of the Forums northern side aisle. The process is mirrored to establish the outer edge of the southern side aisle and results in defining a larger square of 1.618 unit sides.
1.50 /2	**Step Three** – A line is established just inside the North portico edge. A 1.50 technique is performed on the lower left quadrant of the original square generating a vertical line. This line is extended eastward until it intersects the larger square previously created. This intersection will provide the center points for the placement of the Forums two exedras.
1.618 /4	**Step Four** – The sizing of the exedrae is established by defining a circle's diameter from the northern portico edge to the midpoint of the original square's western side using the 1.618 technique of Step Two. The circles are bisected by the outer portico lines with the smaller outer segments remaining as the exedrae.

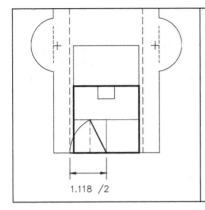

1.118 /2

Step Five – Here, using a 1.118 technique, the innermost part of both side aisles is a colonnade baseline adjacent to steps into the courtyard. The colonnade extends from the west side of the Forum to its east.

Intervening Comment

This concludes the development of the Forum precinct and before I proceed to do the same with the temple I would like to talk about the issue of alternatives. In the process of writing this paper and having it reviewed by various peers, alternative solutions have been proposed. Some of these were improvements and some were not. In all cases, suggested changes need to fit precisely in my ongoing process of development or the critic was expected to propose a complete process of their own.

Even my own process necessitated different steps to accurately represent different published versions of the Forum of Augustus. There are essentially two different published versions of the Forum, the Gismondi-Bauer version as found in the *Lexicon Topographicon* [Steinby 1995], and the Rodico Reif version as found in Favro's Urban Image [Favro 1996]. The Gismondi-Bauer version lacks parts of the Forum precinct but it non-the-less preferred by J.H. Humpfrey, editor of the *Journal of Roman Archaeology*, as the best choice. Between the two images, the primary differences are the length of the Temple of Mars-Altor and the radius/center point of its apse. However, due to its complete representation of the entire precinct as well as the temple, I have decided to present here my step-by-step comparison to the Rodico-Reif representation.

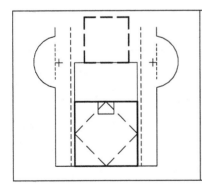

Step Six – Within the original square is placed a square rotated 45°. This new square is rotated again and placed adjacent to and centered on the upper edge of the rectangle derived in Step One. This square defines the temple platform.

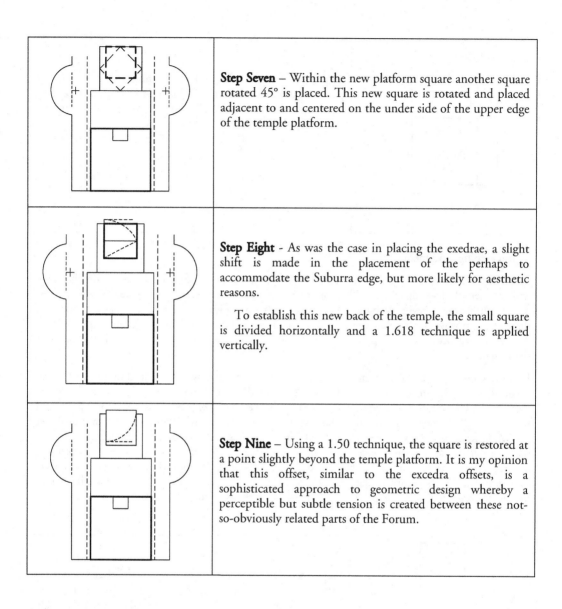

Step Seven – Within the new platform square another square rotated 45° is placed. This new square is rotated and placed adjacent to and centered on the under side of the upper edge of the temple platform.

Step Eight - As was the case in placing the exedrae, a slight shift is made in the placement of the perhaps to accommodate the Suburra edge, but more likely for aesthetic reasons.

To establish this new back of the temple, the small square is divided horizontally and a 1.618 technique is applied vertically.

Step Nine – Using a 1.50 technique, the square is restored at a point slightly beyond the temple platform. It is my opinion that this offset, similar to the excedra offsets, is a sophisticated approach to geometric design whereby a perceptible but subtle tension is created between these not-so-obviously related parts of the Forum.

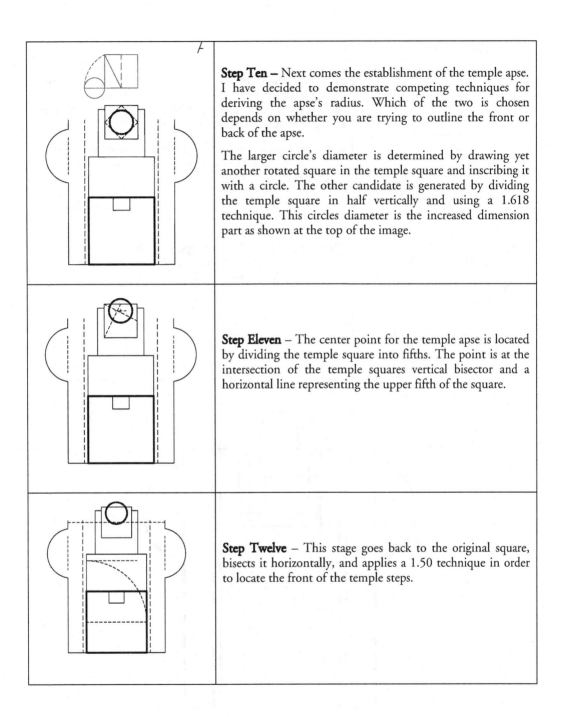

Step Ten – Next comes the establishment of the temple apse. I have decided to demonstrate competing techniques for deriving the apse's radius. Which of the two is chosen depends on whether you are trying to outline the front or back of the apse.

The larger circle's diameter is determined by drawing yet another rotated square in the temple square and inscribing it with a circle. The other candidate is generated by dividing the temple square in half vertically and using a 1.618 technique. This circles diameter is the increased dimension part as shown at the top of the image.

Step Eleven – The center point for the temple apse is located by dividing the temple square into fifths. The point is at the intersection of the temple squares vertical bisector and a horizontal line representing the upper fifth of the square.

Step Twelve – This stage goes back to the original square, bisects it horizontally, and applies a 1.50 technique in order to locate the front of the temple steps.

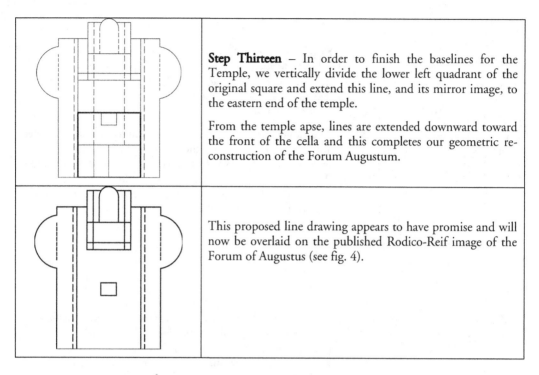

Step Thirteen – In order to finish the baselines for the Temple, we vertically divide the lower left quadrant of the original square and extend this line, and its mirror image, to the eastern end of the temple.

From the temple apse, lines are extended downward toward the front of the cella and this completes our geometric reconstruction of the Forum Augustum.

This proposed line drawing appears to have promise and will now be overlaid on the published Rodico-Reif image of the Forum of Augustus (see fig. 4).

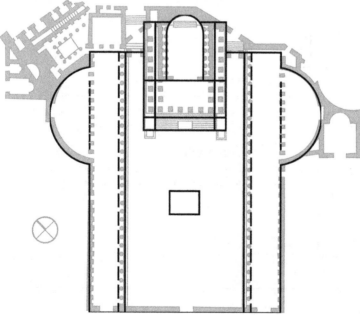

Fig. 4. Comparison of proposed derivation and the Rodico Reif representation

Conclusion

While considerably accurate, there is room to contest this reconstruction at several points. However, I request those tempted to do so refer to the "intervening comments" earlier in this paper.

Select Bibliography

ADAM, Jean Pierre. 1994. *Roman Building: Materials and Techniques*. Bloomington: Indiana University Press.

AURIGEMMA, Salvatore. 1963. *The Baths of Diocletion and the Museo Nazionale Romano*. Rome: Instituto Poligrafico Dello Stato.

BALDWIN, Edward. 1814. *The Pantheon: or Ancient History of the Gods of Greece and Rome*. London: Godwin.

BARTON, Tamsyn S. 1994. *Astrology, Physiognomics, and Medicine under the Roman Empire*. Ann Arbor: University of Michigan Press.

BENARIO, Herbert W. 1980. *Hadriani in the Historia Augusta*. American Philological Association. Scholars Press.

———. "Hadrian", http://www.roman-emperors.org/hadrian.htm.

BIRLEY, Anthony R. 1977. *The Roman Emperior Hadrian*. Northumberland: Barcombe Publications.

———. 1997. *Hadrian: the Restless Emperor*. New York: Routledge.

BLACKWELL, William. 1984. *Geometry in Architecture*. New York: Wiley Interscience.

BOATWRIGHT, Mary Taliaferro. 1987. *Hadrian and the City of Rome*. Princeton NJ: Princeton University Press.

BRUNÉS, Tons. 1967. *Secrets of Ancient Geometry and its use*. Copenhagen: Rhodos.

CARY, Earnest. 1955. *Dio's Roman History*, vol. VIII. London: Harvard University Press.

CONNOLLY, Peter and Hazel DODGE. 1998. *The Ancient City: Life in Classical Athens and Rome*. Oxford: Oxford University Press.

Critchlow, Keith. 1970. *Order in Space: a design source book*. New York: Viking Press.

DE FINE LICHT, Kjeld. *The Rotunda in Rome: A Study of Hadrian's Pantheon*. Copenhagen: Jutland Archeological Society.

DeLAINE, Janet. 1997. *The Baths of Caracalla*. Portsmouth, RI.

DEWING, H.B., ed. 1061. *Procopius Volume VII Buildings*. London: Harvard University Press.

DOCZI, Gyorgy. 1981. *The Power of Limits: Proportional Harmonies in Nature, Art and Architecture*. Boulder: Shambhala, Boulder.

FAVRO, Diane. 1006. *The Urban Image of Augustan Rome*. Cambridge: Cambridge University Press.

FULFORD, Eric. 1994. A Temple Through Time. *Archaeology*, Sept-Oct 1994.

HAMBIDGE, Jay. 1932. *Practical Applications of Dynamic Symmetry*. New Haven: Yale University Press.

HIBBERT, Christopher. 1987. *Rome: the Biography of a City*. London: Penguin Books.

ISAAC, Benjamin. 1990. *The Limits of Empire: the Roman Army in the East*. Oxford: Clarendon Press.

JOHNSON, Anne. 1983. *Roman Forts: of the First and Second Centuries*. New York: St. Martins Press.

JOSEPHUS, Flavius. 93 A.D. *The Jewish War*, III.5-6, "The Roman Army in the First Century CE." http://www.fordham.edu/halsall/ancient/josephus-warb.html

KAHLER, Heinz. 1964. The Pantheon as Sacral Art. *Bucknell Review*.

KENNEDY, David and Derrick Riley. 1990. *Rome's Desert Frontier: From the Air*. Austin: University of Texas Press.

KLEINER, Fred S. 1985. *The Arch of Nero in Rome*. Rome: Giorgio Bretschneider Editore.

KOSTOVSKII, A.N. 1961. *Geometrical Constructions using Compasses Only*. New York: Blaisdell Publishing.

KRAPP, Philip. 1996. *Archaic Roman Religion*, vol. 2. Baltimore: Johns Hopkins.

LAMBERT, Royston. 1984. *Beloved and God: the Story of Hadrian and Antinous*. New York: Viking Press.

LANCIANI, Rodolpho. 1897. Pantheon. In *The Ruins and Excavations of Ancient Rome*.

LAWLOR, Robert. 1982. *Sacred Geometry: philosophy and practice*. New York: Crossroad.

LE BOHEC, Yann. 1994. *The Imperial Roman Army*. New York: Hippocrene Books.

LONG, Charlotte R. 1987. *The Twelve Gods of Greece and Rome*. New York: E.J. Brill.

LUNDY, Miranda. 2001. *Sacred Geometry*. New York: Walker and Company.

MACDONALD, William L. 1965. *The Architecture of the Roman Empire*, vol. 1. New Haven: Yale University Press.

———. 1976. *The Pantheon: Design, Meaning, and Progeny*. Cambridge MA: Harvard University Press.

MACDONALD, William L. and John PINTO. 1995. *Hadrians Villa and its Legacy*. New Haven: Yale University Press.

MACKENDRICK, Paul. 1958. *The Roman Mind at Work*. New York: Van Nostrand.

PALLADIO, Andrea. 1965. *The Four Books of Architecture*, chapter XX of the Pantheon. New York, Dover.

POLYBIUS. 1979. Rise of the Roman Empire. Bk. IV in *The Roman Military System*. F.W. Walbank, ed.; Ian Scott-Kilvert, trans. Penguin Books.

RICHARDSON, L. Jr. 1992. *A New Topographical Dictionary of Ancient Rome*. Baltimore: Johns Hopkins.

Roman-Britain.org. Roman Military Glossary. http://www.Roman-Britain.org.

SCOBIE, Alex. 1990. *Hitler's State Architecture: Hitler and Hadrian's Pantheon*. London: Penn State University Press.

SMITH, E. Baldwin. 1956. *Architectural Symbolism of Imperial Rome and the Middle Ages*. Princeton, NJ: Princeton University Press.

SMOGORZHEVSKII, A.S. 1957. *The Ruler in Geometrical Constructions*. New York: Blaisdell Publishing.

SPERLING, Gert. 1999. *Das Pantheon in Rom*. Neuried: Ars Una

STEINBY, Eva Margareta. 1995. *Lexicon Topographicum Urbis Romae*, vol. 2, D-G. Rome: Edizione Quasar.

SYME, Ronald. 1968. *Ammianus and the Historia Augusta, the Letter of Hadrian*. Oxford: Clarendon Press.

TAGLIAMONTE, Gianluca. 1998. *Baths of Diocletian*. Gregory Bailey, trans. Soprintendenza Archeologica di Roma. Milan: Electa.

TODD, Malcolm. 1972. *The Walls of Rome*. New Jersey: Rowman and Littlefield.

TUCKER, T.G. 1910. *Life in the Roman World of Nero and St. Paul*. New York: Macmillan Co.

VANN, Robert Lindley. 1976. A Study of Roman Construction in Asia Minor. Ph.D. thesis, Rensselaer Polytechnic Institute.

VITRUVIUS. 1960. *Ten Books on Architecture*. Morris Hicky Morgan, trans. 1914. Rpt. New York: Dover.

WEBSTER, Graham. 1985. *The Roman Imperial Army: of the first and second centuries*. New Jersey: Barnes and Noble.

WEYL, Hermann. 1982. *Symmetry*. Princeton NJ: Princeton University Press.

Wilson Jones, Mark. 2000. *Principles of Roman Architecture*. New Haven: Yale University Press.

YEGUL, Fikret. 1992. Baths and Bathing in Classical Antiquity. Cambridge MA: MIT Press.

ZANKER, Paul. 1988. *The Power of Images in the Age of Augustus*. Ann Arbor: University of Michigan Press.

About the author

Marshall is a registered architect in the state of Kentucky and a tenured faculty member of the Purdue Architectural Engineering Technology program at Fort Wayne, Indiana. He is currently the coordinator of the Center for the Built Environment at Indiana-Purdue Fort Wayne (IPFW), which facilitates service projects between the university and the community. Marshall received his Bachelor of Architecture degree from the University of Kentucky and his Masters degree from the Graduate School of Design at Harvard. He is currently researching and writing on the process of geometric construction as it relates to Roman Imperial architecture including the Forum of Augustus, the Baths of Caracalla and the Pantheon.

The Geometer's Angle

Rachel Fletcher | *The Golden Section*

113 Division St.
Great Barrington, MA
01230 USA
rfletch@bcn.net

To Renaissance mathematician Luca Pacioli, it was the Divine Proportion. To German astronomer Johannes Kepler, it was a precious jewel. The only proportion to increase simultaneously by geometric progression and by simple addition, the Golden Section achieves unity among diverse elements in remarkably efficient ways. We explore the Golden Ratio 1:ϕ, also known as the Golden Mean, and its appearance in the regular pentagon and other geometric constructions.

I Introduction

To Renaissance mathematician Luca Pacioli, it was the Divine Proportion. To German astronomer Johannes Kepler it was a precious jewel.[1] The only proportion to increase simultaneously by geometric progression and by simple addition, the Golden Section achieves unity among diverse elements in remarkably efficient ways. Though not without its detractors, its appearance has been observed in nature, design and architecture; from Egyptian pyramids and the Parthenon of ancient Greece to Le Corbusier's Modular system; from sunflowers and daisies in the plant world to spiral shells beneath the seas.

Definition:

When a line is divided into two unequal parts (*a* and *b*) such that the shorter part relates in length to the longer part in the same way as the longer part relates to the whole (*a*:*b* :: *b*:*a* + *b*), the result is a **Golden Section**:[2]

SHORTER:LONGER :: LONGER:SHORTER + LONGER (or WHOLE)

$$a : b :: b : a + b$$

The only ratio capable of generating proportions consisting of just two terms (*a* and *b*), the Golden Section is signified by the Greek letter *phi* or ϕ (= $\sqrt{5}/2$ + 1/2), after the Greek sculptor Phidias, and translates numerically to an incommensurable ratio of 1:1.618034..... or 1:ϕ. The reciprocal, 1/ϕ ($\sqrt{5}/2$ - 1/2), equals 0.618034....

The ϕ number series 1/ϕ^3, 1/ϕ^2, 1/ϕ, 1, ϕ, ϕ^2, ϕ^3... increases simultaneously by geometric proportion (1/ϕ : 1 :: 1 : ϕ) and by simple addition (1/ϕ + 1 = ϕ).

$$1/\phi^3, \ 1/\phi^2, \ 1/\phi, \ 1, \ \phi, \ \phi^2, \ \phi^3...$$

$$.236..., \ .382..., \ .618..., \ 1, \ 1.618..., \ 2.618..., \ 4.236....$$

The Golden Section, ϕ, is also called the Golden Ratio, the Golden Mean and the "extreme and mean" ratio.[3]

1590-5896/06/010067-23 DOI 10.1007/s00004-006-0004-z

II The pentagon

ϕ is an incommensurable number that cannot be stated as a whole number fraction, but appears with absolute precision in the geometry of a pentagon.

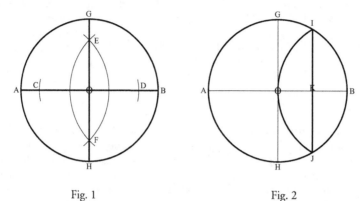

Fig. 1   Fig. 2

- With a compass, draw a circle.

- Draw the horizontal diameter AB through the center of the circle.

- Set the compass at an opening that is slightly smaller than half the radius of the circle. Place the compass point at the center of the circle (point O). Draw arcs that cross the horizontal diameter on the left and right, at points C and D.

- Set the compass at an opening that is slightly larger than before. Place the compass point at C. Draw an arc above and below, as shown.

- With the compass at the same opening, place the compass point at D. Draw an arc above and below, as shown.

- Locate points E and F where the two arcs intersect.

- Draw the line EF through the center of the circle.

- Extend the line EF in both directions to the circumference of the circle (points G and H).

Lines AB and GH locate the horizontal and vertical diameters of the circle (fig. 1).

***

- Locate point B at the right end of the horizontal diameter (AB).

- Place the compass point at B. Draw an arc of radius BO that intersects the circle at points I and J.

- Connect points I and J.

- Locate point K, where the line IJ intersects the horizontal diameter (AB).

Point K divides the radius OB in half (fig. 2).

***

- Connect points K and G.

- Place the compass point at K. Draw an arc of radius KG that intersects the horizontal diameter (AB) at point L (fig. 3).

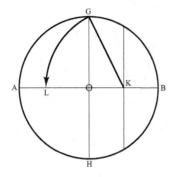

Fig. 3   Fig. 4

- Connect points G and L.
- Place the compass point at G. Draw an arc of radius GL that intersects the circle at point M, as shown (fig. 4).

***

- Connect points G and M.

Line GM locates the side of a regular pentagon inscribed within the circle.

- Place the compass point at M. Draw an arc of radius MG that intersects the circle at point N, as shown.
- Place the compass point at N. Draw an arc of the same radius that intersects the circle at point P.
- Place the compass point at P. Draw an arc of the same radius that intersects the circle at point Q.
- Connect the five points G, M, N, P and Q.

The result is a regular pentagon inscribed within the circle (fig. 5).

***

- Draw the diagonals NG and PG within the pentagon.

If the side (PN) of the pentagon is 1, the diagonal (NG) equals ϕ (= $\sqrt{5}/2 + 1/2$ or 1.618034...).

The side and diagonal of any regular pentagon are in the ratio 1:ϕ (fig. 6).

***

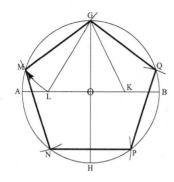

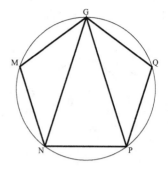

PN : NG :: 1 : Φ

Fig. 5   Fig. 6

Analysis by the Pythagorean Theorum. To understand the pentagon's inherent ϕ proportions, we must consider the ½:1 or 1:2 right triangle that initiates its construction.

- Locate the right triangle GOK.

The side GO coincides with the radius of the original circle and equals 1. The side OK coincides with half the radius and equals ½ .

By the Pythagorean Theorem,

$$GO^2 + OK^2 = KG^2$$
$$1^2 + (½)^2 = (5/4) = KG^2$$
$$KG = \sqrt{5}/2 \quad (fig. 7.)[4]$$

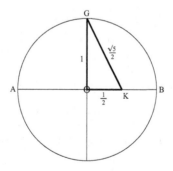

Fig. 7

- Place the compass point at K. Draw an arc of radius KG that intersects the horizontal diameter (AB) at point L, as shown.

$KG = KL = \sqrt{5}/2$

$OL = KL - KO = \sqrt{5}/2 - \frac{1}{2} = 1/\phi$

$BL = KL + BK = \sqrt{5}/2 + \frac{1}{2} = \phi$

$LA = OA - OL = 1 - (\sqrt{5}/2 - \frac{1}{2}) = (3 - \sqrt{5})/2 = 1/\phi^2$

$LA{:}OL :: OL{:}OG :: OG{:}LB$

$1/\phi^2 {:} 1/\phi :: 1/\phi {.} 1 :: 1{:}\phi$

- Locate the right triangle LOG.

The side LO equals $1/\phi$. The side OG equals 1.

By the Pythagorean Theorem,

$LO^2 + OG^2 = GL^2$

$(1/\phi)^2 + (1)^2 = (1/\phi^2 + 1) = GL^2$

$GL = \sqrt{(1/\phi^2 + 1)}$

The hypotenuse GL equals the side (GM) of the pentagon inscribed within the original circle (fig. 8).

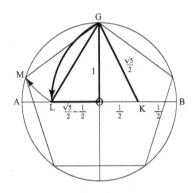

$LA : OL :: OL : OG :: OG : LB$

$\frac{1}{\Phi^2} {:} \frac{1}{\Phi} :: \frac{1}{\Phi} {:} 1 :: 1 : \Phi$

Fig. 8

III Golden ratios in the pentagon

Each segment of a regular pentagonal system relates to the others according to a variable of ϕ.

- Draw the pentagon (GMNPQ) and its five diagonals.

If the side (PN) of the pentagon is 1, the diagonal (NG) equals ϕ.

But if the segment SP equals 1:

$RS = 1/\phi$

$PN = \phi$

$NG = \phi^2$

The segments increase simultaneously by geometric proportion and by simple addition (fig. 9).

RS:SP :: SP:PN :: PN:NG

$1/\phi : 1 :: 1 : \phi :: \phi : \phi^2$

.618...:1 :: 1:1.618... :: 1.618...: 2.618....

At the same time,

RS + SP = PN

$1/\phi + 1 = \phi$

.618... + 1 = 1.618...

SP + PN = NG

$1 + \phi = \phi^2$

1 + 1.618... = 2.618...

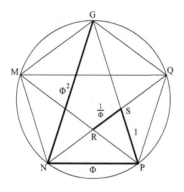

RS : SP :: SP : PN :: PN : NG

$\frac{1}{\Phi} : 1 :: 1 : \Phi :: \Phi : \Phi^2$

RS + SP = PN

$\frac{1}{\Phi} + 1 = \Phi$

Fig. 9

IV The Golden Triangle

Definition:

The **Golden** or **Sublime Triangle** is an isosceles triangle formed by two diagonals and one edge of a regular pentagon. It contains one 36° angle and two 72° angles. The Golden Triangle divides into a 108°-36°-36° isosceles triangle and a reciprocal that is proportionally smaller in the ratio $1 : 1/\phi.^5$

- Draw the pentagon (GMNPQ) and its five diagonals.
- Locate the Golden 36°-72°-72° triangle GPN.
- Locate the line PT, as shown.

Line PT divides the major triangle (GPN) into a 108°-36°-36° triangle (TGP) and a reciprocal (PNT) that is proportionally smaller in the ratio 1 : 1/ϕ.

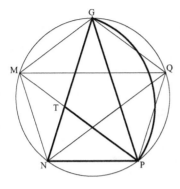 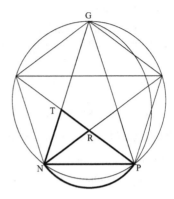

Fig. 10   Fig. 11

- Place the compass point at T. Draw an arc of radius TG that intersects the line GP at points G and P (fig. 10).

***

- Locate the Golden 36°-72°-72° triangle PNT.
- Locate the line NR, as shown.

Line NR divides the major triangle (PNT) into a 108°-36°-36° triangle (RPN) and a reciprocal (NTR) that is proportionally smaller in the ratio 1 : 1/ϕ.

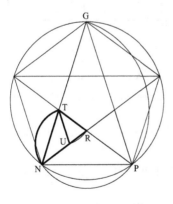 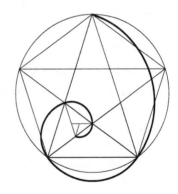

Fig. 12   Fig. 13

- Place the compass point at R. Draw an arc of radius RP that intersects the line PN at points P and N (fig. 11).

***

- Locate the Golden 36°-72°-72° triangle NTR.

- Place the compass point at T. Draw an arc of radius TR that intersects the line RN at point U, as shown.

- Locate the line TU.

- Place the compass point at U. Draw an arc of radius UN that intersects the line NT at points N and T (fig. 12).

***

- Repeat the process continually, as shown.

The result is an equiangular spiral in the ratio 1 : ϕ (fig. 13).

V *How to draw a Golden Mean rectangle*

The Golden Mean proportion may take many forms and expressions. For example, the diagonal of half a square yields a rectangle in the ratio 1:ϕ.

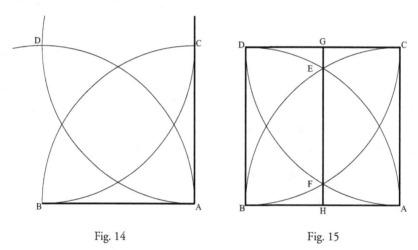

Fig. 14   Fig. 15

- Draw a horizontal baseline AB equal in length to one unit.

- From point A, draw an open-ended line perpendicular to line AB, that is slightly longer in length.

- Place the compass point at A. Draw a quarter-arc of radius AB that intersects the line AB at point B and the open-ended line at point C.

- Place the compass point at B. Draw a quarter-arc (or one slightly longer) of the same radius, as shown.

- Place the compass point at C. Draw a quarter-arc (or one slightly longer) of the same radius, as shown.

- Locate point D, where the two quarter-arcs, taken from points B and C, intersect.

- Place the compass point at D. Draw a quarter-arc of the same radius that intersects the line CA at point C and the line BA at point B (fig. 14).

***

- Connect points B, D, C and A.

The result is a square (BDCA) of side 1.

- Locate points E and F where the quarter-arcs intersect.
- Draw the line EF.
- Extend the line EF in both directions to points G and H on the square.

Line GH divides the square (BDCA) in half (fig. 15).

***

- Connect points H and C.
- Place the compass point at H. Draw an arc of radius HC that intersects the extension of line BA at point I.
- From point I, draw a line perpendicular to line IB that intersects the extension of line DC at point J.

The rectangle (JIBD) that results is in the ratio $1:\phi$.[6]

JI:IB :: $1:\phi$

The major $1:\phi$ rectangle (JIBD) divides into a square (DCAB) and a reciprocal (CJIA) that is proportionally smaller in the ratio $1 : 1/\phi$.[7]

The long side (IB) of the major $1 : \phi$ rectangle (JIBD) equals the sum of the short (CJ) and long (JI) sides of the reciprocal (CJIA) (fig. 16.).

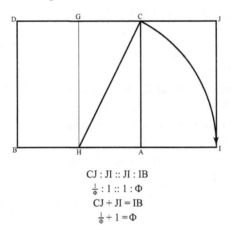

$$CJ : JI :: JI : IB$$
$$\tfrac{1}{\Phi} : 1 :: 1 : \Phi$$
$$CJ + JI = IB$$
$$\tfrac{1}{\Phi} + 1 = \Phi$$

Fig 16

VI The rectangle of whirling squares

- Place the compass point at A. Draw a quarter-arc of radius AI that intersects the line CA at point K.

- Place the compass point at I. Draw a quarter-arc of radius IA that intersects the line JI at point L.
- Connect points K and L.

The line KL divides the major 1:φ rectangle (CJIA) into a square (AKLI) and a reciprocal (KCJL) that is proportionally smaller in the ratio 1 : 1/φ (fig. 17).

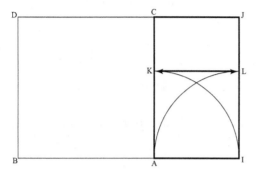
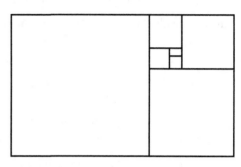

KC : CJ :: CJ : JI :: JI : IB

$\frac{1}{\Phi^2}:\frac{1}{\Phi} :: \frac{1}{\Phi}: 1 :: 1 : \Phi$

Fig. 17   Fig. 18

- Repeat the process continually, as shown.

The result is rectangle of whirling squares (fig. 18).

***

- Place the compass point at C. Draw a quarter-arc of radius CD that intersects the line BI at point A.
- Place the compass point at K. Draw a quarter-arc of radius KA that intersects the line IJ at point L.
- Place the compass point at M. Draw a quarter-arc of radius ML that intersects the line JD at point N.
- Repeat the process continually to reveal a spiral of quarter-arcs whose radii decrease in the ratio 1 : 1/φ.

The quarter-arcs decrease toward a fixed point of origin (the pole or eye), but never touch it (fig. 19).

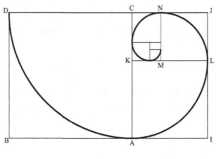

Fig. 19

- Draw the diagonal BJ of the major rectangle JIBD
- Draw the diagonal IC of the reciprocal CJIA.

The diagonals (BJ and IC) intersect at 90° at point O (fig. 20).

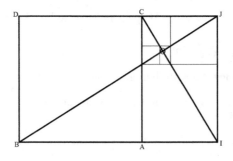

IC : BJ :: 1 : Φ

Fig. 20

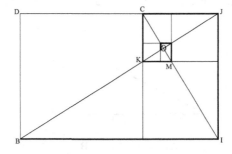

OK : OC :: OC : OJ :: OJ : OI :: OI : OB :: 1 : Φ

MK : KC :: KC : CJ :: CJ : JI :: JI : IB

$\frac{1}{\Phi^3} : \frac{1}{\Phi^2} :: \frac{1}{\Phi^2} : \frac{1}{\Phi} :: \frac{1}{\Phi} : 1 :: 1 : \Phi$

MK + KC = CJ

$\frac{1}{\Phi^3} + \frac{1}{\Phi^2} = \frac{1}{\Phi}$

Fig. 21

- Locate the equiangular spiral of straight-line segments MK, KC, CJ, JI and IB.
- Locate the pole or eye of the spiral at the intersection of the diagonals BJ and IC (point O).
- Locate the spiral's radii vectors OK, OC, OJ, OI and OB.[8]

The radii vectors are separated by equal angles (90°). Their lengths increase in the ratio 1:ϕ. The sum of two adjacent radii vectors equals the length of the next larger vector.

The equiangular spiral BIJCKM decreases in the ratio 1 : 1/ϕ toward a fixed point of origin (the pole at point O), but never touches it (fig. 21).

VII How to divide a line in Golden Section

In previous examples, the ϕ ratio was obtained by adding a new length of 1/ϕ to a line of one unit. In this construction, the line of one unit is divided, in the ratio 1/ϕ^2 : 1/ϕ or 1 : ϕ.

- Draw a baseline (AB) equal to 1.
- From point A, draw a line AC perpendicular to line AB, equal to half the length of AB.
- Connect points C and B.

The result is a right triangle CAB with short and long sides in the ratio ½ :1 or 1:2.

- Place the compass point at C. Draw an arc of radius CA that intersects the hypotenuse BC at point D.
- Place the compass point at B. Draw an arc of radius BD that intersects the long side AB at point E.

Point E divides the side AB in Golden Section

If the line AB is 1, segments AE and EB equal 1/ϕ^2 and 1/ϕ, respectively (fig. 22).[9]

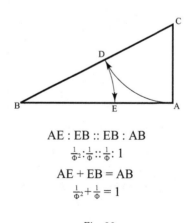

$$AE : EB :: EB : AB$$
$$\tfrac{1}{\phi^2} : \tfrac{1}{\phi} :: \tfrac{1}{\phi} : 1$$
$$AE + EB = AB$$
$$\tfrac{1}{\phi^2} + \tfrac{1}{\phi} = 1$$

Fig. 22

- Place the compass point at B. Draw an arc of radius BE that intersects a line drawn from point B, perpendicular to line BA, at point F.
- From point F, draw a line perpendicular to line FB that intersects the extension of line AC at point G.
- Connect points G, A, B and F.

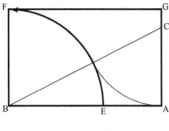

$$GA : AB :: \tfrac{1}{\Phi} : 1$$

Fig. 23

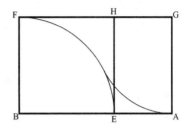

$$HG : GA :: GA : AB$$
$$\tfrac{1}{\Phi^2} : \tfrac{1}{\Phi} :: \tfrac{1}{\Phi} : 1$$
$$HG + GA = AB$$
$$\tfrac{1}{\Phi^2} + \tfrac{1}{\Phi} = 1$$

Fig. 24

The result is a $1/\phi$:1 or 1:ϕ rectangle (GABF) (fig. 23).

***

- From point E, draw a line perpendicular to line BA that intersects the line FG at point H.

The major $1/\phi$: 1 rectangle (GABF) divides into a square (FHEB) and a reciprocal (HGAE) that is proportionally smaller in the ratio 1 : $1/\phi$ (fig. 24).

VIII How to draw a pentagon from a square

In this construction, we draw a regular pentagon from a square of side 1.[10]

- Draw a square (ABCD) of side 1.
- Place the compass point at A. Draw a semicircle of radius AB that intersects the extension of line BA at point E.
- Place the compass point at B. Draw a semicircle of the same radius that intersects the extension of line AB at point F (fig. 25).

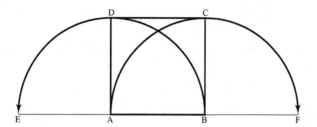

Fig. 25

- Place the compass point at C. Draw a quarter-arc of radius CB that intersects the square at points B and D.

- Place the compass point at D. Draw a quarter-arc of the same radius that intersects the square at points A and C.

- Locate points G and H where the quarter-arcs and semicircles intersect, as shown.

- Draw the line GH.

- Extend the line GH in both directions to points I and J on the square.

Line IJ divides the square (ABCD) in half (fig. 26).

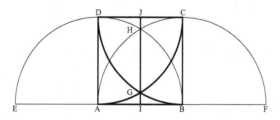

Fig. 26

- Connect points I and C.

- Draw a semicircle of radius IC that intersects the line EF at points K and L (fig. 27).

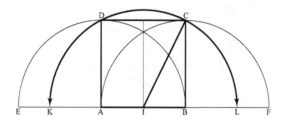

Fig. 27

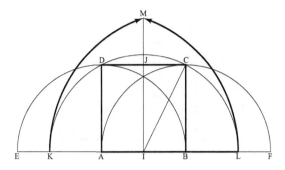

Fig. 28

- Place the compass point at A. Draw an arc of radius AL that intersects the extension of line IJ at point M.

- Place the compass point at B. Draw an arc of the same radius that intersects the extension of line IJ at point M (fig. 28).

**

- Locate points N and O, as shown.

- Connect points A, B, O, M and N.

The result is a regular pentagon whose side of 1 equals the side of the square (ABCD) (fig. 29).

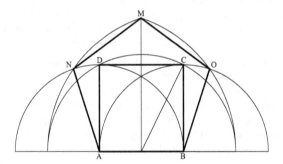

Fig. 29.

Analysis by the Theorum of Thales. The construction for drawing a pentagon from a square is based on the ϕ relationships that result when a square is inscribed within a semicircle.[11] The construction further demonstrates the Theorem of Thales; that any triangle inscribed within a semicircle is right-angled.[12]

- Locate the semicircle drawn on the diameter KL.

- Locate points K, D and L.

- Connect points K, D and L.

The result is a right triangle inscribed within the semicircle.

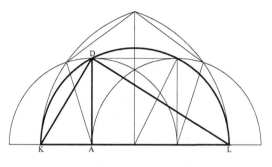

AK : AD :: AD : AL

$\frac{1}{\Phi}$: 1 :: 1 : Φ

Fig. 30

The Theorem of Thales states that within a semicircle, a perpendicular line (DA) drawn from any point (D) along the perimeter to the diameter is the mean proportional or geometric mean of the two line segments (AK and AL) that result on the diameter (fig. 30).

IX History of the Golden Ratio

The "extreme and mean" ratio appears as early as Euclid, if not before, and is recognized as a mathematical principle by art and architectural theorists of the Renaissance such as Leon Battista Alberti, Sebastiano Serlio, Albrecht Dürer and Luca Pacioli.[13] But the origin and history of its actual use in art and architecture are rigorously debated. Opponents are careful to distinguish ϕ as a mathematical principle from its design application. Marcus Frings [2002] and others argue that the "extreme and mean" ratio does not appear in Vitruvius's canon of proportion, and therefore architectural theorists of the Renaissance who rediscovered Vitruvian principles are unlikely to have adopted it. Pacioli's name for ϕ is *Divina proportione*, the title of his 1509 treatise where, in the first book (*Compendium de divina proportione*), he discusses the philosophical and theological aspects of the ratio. But the second book, *Tractato de l'architectura*, a treatise on architecture, does not advocate its use in design practice. Few dispute that the Golden Section has appeared architecturally for aesthetic purposes since the mid-nineteenth century, when Adolf Zeising and Gustav Theodor Fechner introduced it to architectural theory [Frings 2002, 9-20; Herz-Fischler 1998, 149-151, 171-172; March 2001, 85-86; Padovan 1999, 304-308; Scholfield 1958, 98-99].

And yet, claims for the Golden Section have been made in architectural works from prehistory, including Neolithic stone circles, Egyptian pyramids, the Parthenon of ancient Greece, and at least one Palladian villa [Critchlow 1982, 87; Fletcher 1995, 9, 17, 23; Fletcher 2000, 73-85; Hambidge 1924, xvi-xvii, 1, 7]. Since the Renaissance, humanists and builders have published exact geometric constructions in art, architectural and building treatises, including, in the sixteenth century, drawings for a pentagon in Dürer's *Underweysung der messung* (*The Painter's Manual*) and in Serlio's compendium on geometry in *Trattato di architettura* (*On Architecture*). A related construction for a decagon appears in Alberti's fifteenth-century *De re aedificatoria* (*On the Art of Building in Ten Books*). In the eighteenth century, Peter Nicholson, Batty Langley, Sébastien Le Clerc and others illustrated exact constructions in manuals for architects and builders [Dürer 1977: II, 144-146; Serlio 1996: I, 29 (fol. 20); Alberti 1988; VII, 196; Le Clerc 1742, 112-3, 180-1; Langley 1726, 11 and pl. 1, fig. XXX; Nicholson 1809, 14 and pl. 13].

X Symbolism of the pentad

In a previous column, we introduced the numbers 1, 2, 3 and 4 of the tetractys, noting their connection to the archetypes of Monad, Dyad, Triad and Tetrad and the qualities of Unity, Multiplicity, Harmony and Body or Form [Fletcher 2005b, 177]. The Pythagorean tradition of identifying numbers with qualities and values extends to the number "five," or Pentad, whose names include Wedding, Marriage, Justice and Light.[14]

"Five" is a circular number and a spherical number, returning to itself in the last digit when raised to the second and third powers (5 x 5 = 25 and 5 x 5 x 5 = 125). The pentad invokes the circle in another way, as the center point that unites the four cardinal directions. The Pythagoreans associated the pentad with the immutable fifth element of ether that comprehends the other four elements. Likewise the fifth regular solid, or dodecahedron, stands for the zodiac and totality.

One name for the pentad is Marriage because it unites the first distinct "species of numbers"— the triad ("three"), which is the first odd or masculine number, and the dyad ("two"), which is the first even or feminine number. The pentad may symbolize the hierogamy, or sacred marriage, of heaven and earth. Another name for the pentad is Justice because, as the middle of the numbers in the decad (1, 2, 3, 4, 5, 6, 7, 8, 9) it achieves equality and balance [Taylor 1972, 189-191].

In a previous column, we equated the hexad, or number "six," with the great world or macrocosm of the universe [Fletcher 2005a, 143]. In similar fashion, the pentad symbolizes the microcosm of humanity or man. The pentagon may represent the human figure with head upright and arms and legs outstretched, while "five" is often associated with the digits of the human hand. Karl Menninger notes that the connection between the number "five" and the "fingers" and "hand" is cross-cultural. The Gothic for "five" is *fimf* and is likely related to the Gothic for "finger," which is *figgrs*. The Slavic for "five" is *pētj* and is likely related to the Slavic for "fist," which is *pěsti*. The Egyptian word for "five" is the same as for "hand" [Menninger 1977, 148-149, Simpson 1989].

XI The Golden Ratio in nature and human body

The Golden Ratio is often linked with the growth of living organisms and has been observed in numerous living forms, including the human anatomy. Allowing for unique and individual differences, the overall proportions of the human face generally conform to a Golden Mean rectangle, while the fingers divide at the joints in ϕ progression (1, ϕ, ϕ^2, ϕ^3...).

Some believe that the tradition of rendering the human body according to ϕ dates from ancient Egypt through modern times with Le Corbusier's Modulor system.[15] Generally speaking, at birth, the navel divides an infant's total height in half. As the infant develops, the navel appears to "rise," eventually marking a Golden Section in the adult male figure. Meanwhile, the half division "lowers" to the place of the sexual organs. But the location of the groin can produce a new ϕ division, when the arms are raised directly overhead.

From these two ϕ divisions—of one's height from head to toe, at the navel; and of one's height with arms directly overhead, at the groin—Le Corbusier derives two intertwined "red" and "blue" scales that comprise the Modulor. The system develops from a square, whose side of 1 equals the height from the floor to a person's navel; and from a double square, whose long side of 2 equals the person's total height with arms raised overhead. The "red" series develops by adding a Golden Section ($1/\phi$) to the square of side 1, locating the top of the head. The "blue" series develops by subtracting a Golden Section ($1/\phi$) from the long length of the double square, locating the groin. The Modulor is expressed in whole numbers and may be adapted to a person of any height or applied to any design situation [Le Corbusier 1980, 50-58, 63-67] (fig. 31).

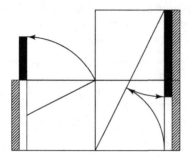

Fig. 31. Le Corbusier's Modulor

XII *The Fibonacci number series*

The thirteenth-century Italian mathematician Leonardo of Pisa, who is also called Fibonacci, popularized the Hindu–Arabic decimal system first introduced to the West by al-Khwarizmi. But Fibonacci is best known for the whole number series that bears his name. Fibonacci numbers simulate a true ϕ progression, increasing by geometric proportion and simple addition, simultaneously.[16]

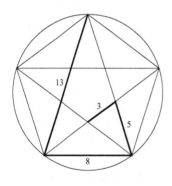

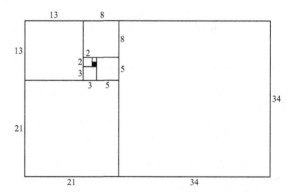

Fig. 32   Fig. 33

The Fibonacci number series reads: 1, 1, 2, 3, 5, 8, 13, 21, 34, 55, 89, 144....

Similar to a true ϕ progression, each number is the sum of the preceding two: 1+1=2; 1+ 2=3; 2+3=5; 3+5=8; 5+8=13... .

At the same time, each successive ratio of adjacent numbers oscillates first above, then below the true value of ϕ (1.618034…).

13/8 (1.62500) is greater than ϕ.
21/13 (1.61538) is less than ϕ.

34/21(1.61904) is greater than ϕ, but closer in value.

55/34 (1.61764) is less than ϕ, and closer still.

The larger the numbers, the more accurately the ratios they form approximate the true value of ϕ. Thus, Fibonacci numbers may represent the segments of a regular pentagonal system (fig. 32) or a rectangle of whirling squares (fig. 33).[17]

Fibonacci numbers can only approximate the incommensurable value of ϕ, but Theodore Cook observes that a double Fibonacci series expresses an exact ϕ progression [Cook 1979, 420]:

$$\phi^2 = 1 + \phi$$
$$\phi^3 = 1 + 2\phi$$
$$\phi^4 = 2 + 3\phi$$
$$\phi^5 = 3 + 5\phi$$
$$\phi^6 = 5 + 8\phi$$

Fibonacci numbers regularly appear in the natural world, as in the display of petals about the center of flowers. The primrose has 5 petals; the ragwort, 13 petals; the daisy, 21 or 34 petals; and the Michelmas daisy, 55 or 89 petals, all Fibonacci numbers.

The seeds of a sunflower arrange about the center in a pattern of opposing sets of curves. In some varieties, 34 long curves spiral clockwise, while 55 short curves spiral in the opposite direction. In others, 55 and 89 curves spiral to the right and left, respectively. When the pattern reaches its spatial limit, a new arrangement of opposing curves may appear, but it, too, follows the Fibonacci sequence. Thus, a sunflower that begins with 34 and 55 curves about the center may change to a new arrangement of 55 and 89.

XIII Application: common grass-of-Parnassus (Parnassia palustris)

A photograph by Karl Blossfeldt presents a common grass-of-Parnassus (*Parnassia palustris*) with the outer pieces removed, leaving the inner parts of the flower to display a complex pattern of five-fold symmetry (fig. 34).

Fig. 34

No single flower conforms exactly to an idealized geometric pattern, but the common grass-of-Parnassus expresses the dynamic symmetry of pentagons with remarkable precision (fig. 35a-d).[18] Whether such geometry is inherent in nature or merely the product of human perception, it reaches beyond the rigid perfection of universal principles to rich and subtle methods of applying abstract rules.

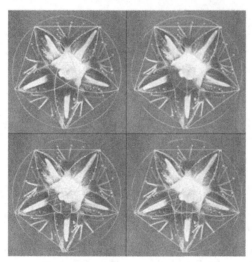

Fig. 35a-d. Photograph: © 2005 Karl Blossfeldt Archiv / Ann u. Jürgen Wilde, Köln / Artists Rights Society (ARS), New York. Geometric overlays by Rachel Fletcher. *Reproduction, including downloading of Karl Blossfeldt's wordks is prohibited by copyright laws and international conventions without the express written permission of Artists Rights Society (ARS), New York.*

Notes

1. Kepler's *Mysterium Cosmographicum* states that "there are two treasure houses of geometry: one, the ratio of the hypotenuse in a right-angled triangle to the sides [the Pythagorean Theorem], and the other, the line divided in the mean and extreme ratio…. The former…can rightly be compared to a mass of gold: the second, on proportional division, can be called a jewel" [Kepler 1981, 133, 143].

2. If the whole is equal to 1, the proportion translates to $[1/\phi^2 : 1/\phi :: 1/\phi : (1/\phi^2 + 1/\phi)$ or 1]. The Golden Section may also express the relationship between a whole line (a) and its longer part (b), such that the whole relates in length to the longer part in the same way as the longer part relates to the whole minus the longer part ($a : b :: b : a - b$). If the whole is equal to 1, the proportion translates to $[1 : 1/\phi :: 1/\phi : (1 - 1/\phi)$ or $\phi^2]$.

3. For a full account of names for the Golden Section, see [Herz-Fischler 1998, 164-170].

4. For more on the Pythagorean Theorem, see [Fletcher 2005b, 152-154].

5. The reciprocal of a major triangle is a figure similar in shape, but smaller in size, such that the short side of the major triangle equals the long side of the reciprocal.

6. Note the similarity between this construction and that of the pentagon. If the side of the square (DCAB) is 1, the diagonal HC of the half -square is $\sqrt{5}/2$. HC = HI = $\sqrt{5}/2$. BI = BH + HI = ½ + $\sqrt{5}/2$ = ϕ. AI = HI – HA = $\sqrt{5}/2$ - ½ = $1/\phi$.

7. The reciprocal of a major rectangle is a figure similar in shape, but smaller in size, such that the short side of the major rectangle equals the long side of the reciprocal. The diagonal of the

reciprocal and the diagonal of the major rectangle intersect at right angles [Hambidge 1967, 30, 131]; see [Fletcher 2004, 103].

8. The radius vector is the variable line segment drawn to a curve or spiral from a fixed point of origin (the pole or eye) [Simpson 1989]; see also [Fletcher 2004, 105].

9. Note the similarity between this construction and those of the pentagon and the $1 : \phi$ rectangle. If the short and long sides of the right triangle CAB are ½ and 1, respectively, the hypotenuse BC equals $\sqrt{5}/2$. BD=BC–DC=$\sqrt{5}/2$–½=$1/\phi$. BD=BE. AB=1. AE=AB–EB=1–$1/\phi$=$1/\phi^2$.

10. This construction appears in [Ghyka 1977, 35].

11. In fig. 27, if the side of the square (ABCD) is 1, the diagonal IC of the half -square is $\sqrt{5}/2$. IC = IL = IK. AI = ½. AL = AI + IL = ½ + $\sqrt{5}/2$ = ϕ. AK=IK–IA=$\sqrt{5}/2$–½=$1/\phi$. Therefore, AK = $1/\phi$, AD = 1, and AL = ϕ.

12. For more on the Theorem of Thales, see [Fletcher 2004, 107].

13. One Euclidean construction divides a line into "extreme and mean ratio," such that the short segment is in ratio to the longer as the longer is in ratio to the whole (or $1 : \phi$) [1956, II: 267 (bk. VI, prop. 30); II: 188 (bk. VI, def. 3)]. Book XIII describes the five regular Platonic solids inscribed within a sphere. Of the five, the icosahedron and the dodecahedron involve the pentagon and its inherent "extreme and mean" proportions [1956, III: 453 (bk. XIII, prop. 8)]. Thomas Heath traces Euclids's division of a line in "extreme and mean" ratio to the early Pythagoreans from whom, Heath assumes, an exact construction of a regular pentagon evolved [Heath 1981, 168.] Keith Critchlow observes a prehistoric application of the ϕ ratio in carvings on a Neolithic sphere housed in Edinburgh, which display the symmetry of a dodecahedron [Critchlow 1982, 149].

14. Of the connection between the pentad and light, the modern Platonist Thomas Taylor says that light is the consequence of circular motion, following the four processes of length, breadth, depth and the "sameness" of the sphere itself [Taylor 1972, 188-189]. Also, Apollo, the god of light, personifies five qualities—omnipotence, omniscience, omnipresence, eternity and unity [Cooper 1978, 116].

15. R. A. Schwaller de Lubicz advocates the Egyptian "tradition" of dividing the human figure in Golden Section at the navel, with the caveat that a small portion of the crown is subtracted from the total height [Schwaller de Lubicz 1998, 313, 325-326, 341-343].

16. Fibonacci's main work, *Liber abaci*, uses the number series 1, 1, 2, 3, 5, 8, 13... to calculate the month-to-month progeny of a single pair of rabbits, such that each pair produces a new pair every month [Huntley 1970, 158-159].

17. To construct the rectangle of whirling squares, begin with a square of side 1. Place a new square of side 1 adjacent to it. The result is a double square. Place a new square of side 2 adjacent to the long side of the double square. The result is a 2 x 3 rectangle. Place a new square of side 3 adjacent to the long side of the 2 x 3 rectangle. The result is a 3 x 5 rectangle. Repeat the process, indefinitely.

18. The first step of the analysis (fig. 35a) replicates Dürer's construction for a five-pointed star, suggesting his appreciation of ϕ as a dynamic proportional system [Dürer 1977: II, 155-156].

References

ALBERTI, Leon Battista. 1988. *On the Art of Building in Ten Books*. Translated by Joseph Rykwert, Neil Leach, and Robert Tavernor. Cambridge: MIT Press.

COOK, Theodore Andrea. 1979. *The Curves of Life*. 1914. Reprint. New York: Dover Publications.

COOPER, J. C. 1978. *An Illustrated Encyclopedia of Traditional Symbols*. London: Thames and Hudson.

CRITCHLOW, Keith. 1982. *Time Stands Still: New Light on Megalithic Science*. New York: St. Martin's Press.

DÜRER, Albrecht. 1977. *The Painter's Manual: A Manual of Measurement of Lines, Areas, and Solids....* Walter L. Strauss, ed. and trans. of the1525 edition. New York: Abaris Books.

EUCLID. 1956. *The Thirteen Books of Euclid's Elements*. Thomas L. Heath, ed. and trans. Vols. I-III. New York: Dover.

FLETCHER, Rachel. 1995. *Harmony by Design: The Golden Mean as a Design Tool*. Exhibition catalog for "Harmony by Design: The Golden Mean." New Paltz, New York: Beverly Russell Enterprises.

————. 2000. Golden Proportions in a Great House: Palladio's Villa Emo. Pp. 73-85 in *Nexus III: Architecture and Mathematics*, K. Williams, ed. Pisa: Pacini Editore.

————. 2001. Palladio's Villa Emo: The Golden Proportion Hypothesis Defended. *Nexus Network Journal* **3**, 2 (Summer-Autumn 2001): 105-112. http://www.nexusjournal.com/Fletcher.html.

————. 2004. Musings on the Vesica Piscis. *Nexus Network Journal* **6**, 2 (Autumn 2004): 95-110. http://www.nexusjournal.com/GA-v6.2.html

————. 2005a. SIX + ONE. *Nexus Network Journal* **7**, 1 (Spring 2005): 141-160. http://www.nexusjournal.com/GAv7n2.html

————. 2005b. The Square. *Nexus Network Journal* **7**, 2 (Autumn 2005): 141-185. http://www.nexusjournal.com/GAv7n2.html

FRINGS, Marcus. 2002. The Golden Section in Architectural Theory. *Nexus Network Journal* **4**, 1 (Winter 2002): 9-32. http://www.nexusjournal.com/Frings.html

GHYKA, Matila. 1977. *The Geometry of Art and Life*. 1946. Reprint, New York: Dover Publications.

HAMBIDGE, Jay. 1924. *The Parthenon and Other Greek Temples: Their Dynamic Symmetry*. New Haven: Yale University Press.

HEATH, Thomas. 1981. *A History of Greek Mathematics*. Vol. I. 1921. Reprint. New York: Dover Publications.

HERZ-FISCHLER, Roger. 1998. *A Mathematical History of the Golden Number*. New York: Dover.

HUNTLEY, H. E. 1970. *The Divine Proportion: A Study in Mathematical Beauty*. New York: Dover.

KEPLER, Johannes. 1981. *Mysterium Cosmographicum: The Secret of the Universe*. Trans. A. M. Duncan of the1596 edition. New York: Abaris Books.

LANGLEY, Batty. 1726. *Practical Geometry Applied to the Useful Arts of Building, Surveying, Gardening, and Mensuration...and Set to View in Four Parts....* London: Printed for W. & J. Innys, J. Osborn and T. Longman, B. Lintot [etc.].

LE CLERC, Sébastien. 1742. *Practical Geometry: or, A New and Easy Method of Treating that Art*. Translated from the French. London: T. Bowles and J. Bowles.

LE CORBUSIER (Charles Edouard Jeanneret). 1980. *Modulor I and II*. Trans. Peter de Francia and Robert Anna Bostock of the1948 and 1955 editions. Cambridge: Harvard University Press.

LIDDELL, Henry George and Robert Scott, eds. 1940. *A Greek-English Lexicon*. Henry Stuart Jones, rev. Oxford: Clarendon Press. Perseus Digital Library Project. Gregory R. Crane, ed. Medford, MA: Tufts University, 2005. http://www.perseus.tufts.edu

LIVIO, Mario. 2002. *The Golden Ratio: The Story of φ, the World's Most Astonishing Number*. New York: Broadway Books.

MARCH, Lionel. 1998. *Architectonics of Humanism: Essays on Number in Architecture*. London: Academy Editions.

————. 2001. Palladio's Villa Emo: The Golden Proportion Hypothesis Rebutted, *Nexus Network Journal* **3**, 4 (Summer-Autumn 2001): 85-104. http://www.nexusjournal.com/March.html

MENNINGER, Karl. 1977. *Number Words and Number Symbols: A Cultural History of Numbers*. Trans. Paul Broneer. Cambridge: The M.I.T. Press.

NICHOLSON, Peter. 1809. *The Principles of Architecture, Containing the Fundamental Rules of the Art, in Geometry, Arithmetic, and Mensuration*. Vols. I-III. London: J. Barfield and T. Gardiner.

PADOVAN, Richard. 1999. *Proportion: Science, Philosophy, Architecture*. London: E & FN Spon.

SCHOLFIELD, P. H. 1958. *The Theory of Proportion in Architecture*. Cambridge: Cambridge University Press.

SCHWALLER DE LUBICZ, R. A. 1998. *The Temple of Man: Apet of the South at Luxor*. Trans. Deborah Lawlor and Robert Lawlor of the 1957 edition. Rochester, Vt: Inner Traditions.

SERLIO, Sebastiano. 1996. *Sebastiano Serlio on Architecture*. Vol. I. Books I-V of *Tutte l'Opere D'Architettura et Prospetiva*. Trans. Vaughan Hart and Peter Hicks of the 1545 (I, II), 1544 (III, IV) and 1547 (V) editions. New Haven: Yale University Press.

SIMPSON, John and Edmund WEINER, eds. 1989. *The Oxford English Dictionary*. 2nd ed. OED Online. Oxford: Oxford University Press. 2004. http://www.oed.com/

TAYLOR, Thomas. 1972. *The Theoretic Arithmetic of the Pythagoreans*. 1816. Reprint. New York: Samuel Weiser.

About the geometer

Rachel Fletcher is a theatre designer and geometer living in Massachusetts, with degrees from Hofstra University, SUNY Albany and Humboldt State University. She is the creator/curator of two museum exhibits on geometry, "Infinite Measure" and "Design By Nature". She is the co-curator of the exhibit "Harmony by Design: The Golden Mean" and author of its exhibition catalog. In conjunction with these exhibits, which have traveled to Chicago, Washington, and New York, she teaches geometry and proportion to design practitioners. She is an adjunct professor at the New York School of Interior Design. Her essays have appeared in numerous books and journals, including "Design Spirit", "Parabola", and "The Power of Place". She is the founding director of Housatonic River Walk in Great Barrington, Massachusetts, and is currently directing the creation of an African American Heritage Trail in the Upper Housatonic Valley of Connecticut and Massachusetts.

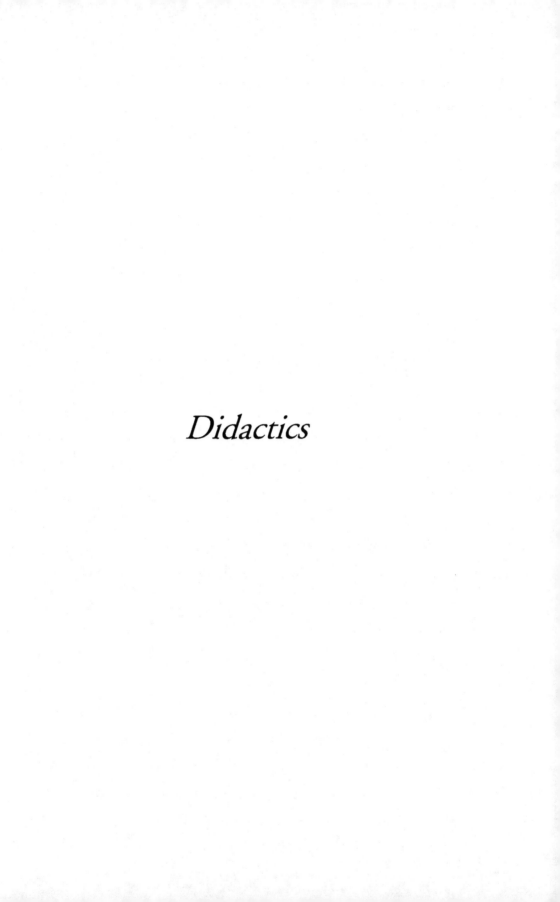

Didactics

Igor M. Verner
Sarah Maor

Department of Education
in Technology & Science
Technion – Israel Institute
of Technology
Haifa, 32000 Israel
ttrigor@tx.technion.ac.il

Mathematical Mode of Thought
in Architecture Design Education: A case study

Integrating mathematics and architecture design curricula has resulted in a positive change in students' abilities to apply mathematics to architectural design. The authors developed the first-year calculus-with-applications course based on the Realistic Mathematics Education approach. In order to encourage students to use mathematics in design projects, the integration of mathematics and architecture education was continued by developing and evaluating the second-year Mathematical Aspects in Architectural Design course based on the Mathematics as a Service Subject approach. The paper considers three directions of geometrical complexity studied in the course with a focus on the process of project-based learning of curved surfaces.

Introduction

Architectural design is treated as a purposeful reflection-in-action process of creating a structure which fits building, material and topographic standards, funding and time limitations, and requirements of culture, aesthetics, environment and proportions. The structure should also match optimization criteria with respect to shape, stability, energy and resources consumption [Lewis 1998, Gilbert 1999, Unwin 1997, Burt 1996]. To answer these professional requirements, in their design practice the architecture program graduates should develop conceptual ideas and implement them in material products, demonstrating deep understanding of cultural, social, technological and management aspects of the project [Gilbert 1999]. These professional skills lean on prerequisite knowledge in art, science and technology [Kappraff 1991].

The value of mathematical thinking in architecture has been emphasized in recent research, particularly in geometrical analyses, formal descriptions of architectural concepts and symbols, and engineering aspects of design [Luhur 1999, Williams 1998]. Architectural educators are calling for enhanced learning programs for mathematics in architectural education by revising the goals, content, and teaching/learning strategies [Salingaros 1999, Williams 1998, Luhur 1999, Consiglieri and Consiglieri 2003, Pedemonte 2001]. Educational research is currently required in order to accommodate didactical approaches to mathematics in architecture education.

The two main approaches to teaching mathematics in various contexts are "Realistic Mathematics Education" (RME) and "Mathematics as a Service Subject" (MSS). In RME, the mathematics curriculum integrates various context problems where the problem situation is experientially real to the student [Gravemeijer and Doormen 1999]. The MSS approach considers mathematics as part of professional education and focuses on mathematical competence required for professional practice [Pollak 1988]. This includes the capability to apply mathematics to design processes such as geometrical design in architecture [Houson et al. 1988, 8].

This paper considers an ongoing study which utilizes the RME and MSS approaches to developing an applications-motivated mathematics curriculum in one of the architecture colleges in Israel. At the first stage we developed a calculus course, based on the RME approach, as part of the first-year mathematics curriculum [Verner and Maor 2003]. The two-year follow-up indicated the positive effect of integrating applications on motivation, understanding, creativity and interest in mathematics. However, from the analysis of fifty-two graduate design projects of students who took the first year RME-based mathematics course, we found that the students did not apply to their architecture design projects in any considerable way the mathematical knowledge acquired in the course.

This situation motivated us to continue the study and develop a "Mathematical Aspects in Architectural Design" (MAAD) course based on the MSS approach. This paper presents architectural design assignments and related mathematical models and activities in the course.

The MAAD course is given in the second college year. It relies on the first year mathematics course [Verner and Maor 2003] and offers mathematical learning as part of hands-on practice in architecture design studio. The course focuses on the analysis of geometrical forms through experiential learning activities.

Geometrical forms

In structuring the course we follow the principles of classification of geometrical forms in architecture education. Consiglieri and Consiglieri [2003] proposed to concentrate mathematics in the architecture curriculum on composing the variety of complex architectonic objects from elementary geometrical forms. Salingaros [2000] formulated the following eight principles of geometrical complexity of urban forms:

- Couplings – connecting elements on the same scale to form a module;
- Diversity – creating couples from different elements;
- Boundaries – connecting modules by their boundaries and not by internal elements;
- Forces – shaping an object by force loading;
- Organization – shaping a structure by superposition of forces;
- Hierarchy – assembling components of various scales from small to large;
- Interdependence – assembling depends on components' properties but not vice versa;
- Decomposition – identifying and analyzing different types of units included in a form.

Grounded in these principles, our study deals with three directions of complexity in geometrical objects for architectural design:

1. **Arranging regular shapes to cover the plane (tessellations).** Boles and Newman [1990] developed a curriculum that studied plane tessellations arranged by basic geometrical shapes with focus on proportions and symmetry. Applications from Fibonacci numbers and the golden section to designing tessellations were emphasized. Frederickson [1997] studied geometrical dissections of figures into pieces and their rearrangements to form other figures using two methods: examining a shape as element of the module, and examining a vertex as a connection of elements. Ranucci [1974] studied mathematical ideas and procedures of tessellation design implemented in Escher's artworks.

2. **Bending bars and flat plates to form curve lines and surfaces (deformations).** Hanaor [1998] introduces a course "The Theory of Structures" which focuses on "the close link between form and structure, between geometry and the flow of forces in the structure". He points out that distributed loads on straight bars and planar surfaces affect bending and shape (deformation) which can be described by different mathematical functions.

The curved surfaces defined by these functions are used to minimize deformation of structures under distributed loads and to express aesthetic principles.

The reciprocal connection between form and construction characterizes Gaudi's approach to architectural design [Alsina and Gomes-Serrano, 2002]. Gaudi systematically applied mechanical modelling to create geometrical forms and to examine their properties. His experimental tools included photographic workshops, plaster models, mirrors, bells, moveable ceilings and other models. He conducted experimental research in order to come up with geometrical solutions which were optimal from the construction viewpoint. He also created 3D surfaces such as paraboloids, helicoids and conoids by moving generator profiles "in a dynamic manner" [Alsina 2002, 89].

3. **Intersecting solids (constructions).** Burt [1996] examined integrating and subdividing space by different types of polyhedral elements. He emphasized that this design method can provide efficient architectural solutions.

Alsina [2002, 119-126] considered the design of complex three-dimensional forms by intersecting various geometrical forms. He showed how Gaudi used these forms in his creations in order to achieve functional purposes, such as the light-shining effects or symbolic expressions.

Case study framework

The Mathematical Aspects in Architectural Design (MAAD) course has been implemented in one of Israeli colleges as part of the architecture program that certifies graduates as practicing architects. This second-year course relied on the first-year mathematics course [Verner and Maor 2003] and offered mathematical learning as part of hands-on practice in architecture design studio.

Following Schoen [1988] we consider the design studio as an experiential learning environment that represents real world practice and involves students in learning by doing, knowing-in-action, and reflection-in-action experience. Using the studio as an authentic environment for architectural design education, the MAAD course offers design assignments which require self-directed mathematical learning. This includes inquiring mathematical aspects of architectural problems, studying new mathematical concepts on a need-to-know basis, and applying them to geometrical design.

The MAAD course consists of three parts corresponding to the three directions of geometrical complexity introduced in the previous section of this paper. Each part of the course includes the following components:

- Mathematical concepts and methods with connections to architecture;

- Practice in solving mathematical problems;

- Design projects.

The 56-hour MAAD course outline is presented in Table 1. The first column includes the three course subjects (tessellations, curved surfaces, and solids intersections). The second column details the above mentioned components for each of the course subjects. The third column contains instructional goals which directed mathematical learning in each of the subjects. The fourth column describes learning activities towards achieving the objectives.

Course subjects	Components	Instructional goals	Learning activities
1. Tessellations	A. Mathematical concepts of tessellations design (4 hours)	Understanding the rules of harmonic dimensions and their use in art, architecture and music	Seminar presentations on golden section, Fibonacci sequence, logarithmic spiral, harmonic properties and applications.
	B. Practice in solving mathematical problems related to tessellations (4 hours)	Acquiring basic skills in analysis of proportions, symmetry, harmonic dimensions, and drawing tessellations	Drawing logarithmic spirals, analyzing basic geometrical figures and their possible combinations, and drawing tessellation fragments.
	C. Tessellation design project (8 hours)	Acquiring experience in tessellation design, making use of modularity, differentiation, proportion and harmony	Designing a flat tessellation of a given floor surface as a combination of modules inspired by a certain metaphoric subject
2. Curved surfaces	A. Algebraic surfaces used in constructions (4 hours)	Identifying types of mathematical surfaces utilized in roof design of existing gas stations and other constructions	Seminar presentations on roof surfaces of existing constructions such as the Sarger surface, hypar, and ellipsoid.
	B. Practice in drawing algebraic surfaces (6 hours)	Developing skills of drawing surfaces through calculating their parameters and coordinates	Exercises of drawing algebraic surfaces (plane, sphere, ellipsoid, cylinder, elliptic hyperboloid, hyperbolic paraboloid) and calculating their volumes and areas.
	C. Gas station design project (10 hours)	Application of calculus to defining complex roof shapes for large span solutions	Designing a gas station with a curved roof which answers the functionality, stability, and constructive efficiency criteria. Building a physical model which accurately represents the roof dimensions.
3. Solids / intersections	A. Algebraic solids and intersections (4 hours)	Defining algebraic solids, their features, parameters and intersections, in application to structures design	Seminar presentations on solid intersections (cylinder with oblique cone and pyramid, sphere with ellipsoid and straight cone) in existing constructions.
	B. Practice in calculating parameters of solids and intersections (6 hours)	Developing skills of mathematical analysis of solids and intersections	Drawing algebraic solids, their sections and involutes (?). Calculating volumes, surface areas, and involutes
	C. Solids and intersections project (10 hours)	Acquiring the skill of analysis of composed structures using analytic geometry	Decomposing an existing architectural structure into solids and analyzing their intersections. Building a physical model of the structure

Table 1. The MAAD course outline

In this paper we will focus on the second course subject, related to surfaces of structures. In the study of curved surfaces, as in the two other subjects, the mathematics concepts were introduced in the form of student seminars. Students were guided with respect to seminar contents, references, and presentation procedures. Each of the seminars was given by a number of students and included definitions of mathematical concepts and their applications in architecture. Hands-on activities and discussions were encouraged. Seminars given by the students dealt with curved surfaces in public buildings. Two examples of the seminars are gas stations with curved roofs, and Sarger surface analysis in the "Le Marche de Royan" building.

The practice sections of the course focused on specific mathematical skills related to the subjects. In addition to exercises of drawing algebraic surfaces the students were required to calculate their parameters and coordinates.

The study of curved surfaces in the course was culminated by a design project assignment formulated as follows:

> *Design a plan and top covering of a gas station. Start from a zero level plan including access roads, parking, pumps, cars washing, coffee shop, and an office. Design a top covering for the pumps area, or the roof of the coffee shop and office building. Find a design solution that addresses the stability, functionality, constructive efficiency, complexity and aesthetics criteria.*

The project stages are:

- Identifying the project data (place, design guidelines for gas stations, dimensions, and prototypes);

- Developing an architectural programme;

- Defining design factors relevant to the project;

- Generating alternative solutions;

- Analyzing alternatives and selecting the solution;

- Producing drawings, calculations, and a physical model.

The projects evaluation is based on the following criteria:

- design criteria: constructive efficiency (8%), aesthetics aspect (8%), architectural functionality (20%), program quality (8%);

- mathematics criteria: gas station dimensions calculation (8%), parametric analysis of surfaces (8%), roof model calculation (8%), building the physical model (8%), precision of model and calculation (8%), mathematical model analysis (8%), geometrical complexity of solutions (8%).

The projects were performed as individual assignments. The course meetings during the five-week period of the project were dedicated to guidance, studio discussions, and progress reports.

Methodology

The first step of the study was examining past senior projects (N=52) with regard to the mathematics applications. It revealed the following features:

1. All the students used proportions in designing spaces, areas and lines, but did not use the golden section or any other irrational ratio.

2. In all the projects, contours were described only by line and circular segments. Even if the students needed to use more complicated lines, they described them by circular segments. Only one of the students used a higher order curve, namely a logarithmic spiral, working under a mathematics teacher's guidance. A few of the students (9.6%) inscribed curved surfaces inside the structure but defined them graphically and not analytically. Students avoided trigonometric calculations of constructive parameters by using computer drawing operations. This method required them to draw additional sections of structure elements, which caused architectural mistakes in measuring distances and angles. Also, no experience was found in the analysis of special points, while conjugations of curves were performed graphically.

3. The final solution was not based on the mathematical optimization analysis. In fact, this analysis was not required by the assignment, which did not specify dimensional and cost limitations and criteria.

4. All design operations related to lengths, areas, and 3D geometry were performed by means of software tools. Lack of analytical evaluation of results caused mistakes in calculating geometrical parameters in the projects.

The insufficient application of mathematical methods in the senior projects indicated the need of an additional course which teaches the applications of mathematics to architectural design. The new MAAD course was implemented in the college in the 2003-2004 academic year by one of the present authors (Sarah Maor) and attended by twenty-six second-year students. The goal of the course's follow-up was to examine mathematical learning in the architectural design studio. The study focused on the following questions:

- What are the features of mathematical learning in the studio environment?

- What is the effect of the proposed environment on learning mathematical concepts and methods?

The study used qualitative and quantitative methods, which analyzed architecture design experience and observed learning behaviour within the context of design studio [Schoen 1988]. Different tools were used for answering the research questions. Data on the features of mathematical learning were gathered from:

- Interviews with experts. Two experienced practicing architects considered their professional activities and relevant mathematical concepts throughout the design stages. From these considerations we derived ideas of the design activities in the three course projects, of the design education features, and of mathematical needs in design.

- Architectural design and mathematics education literature. Relevant teaching methods using context, visualization, heuristic and intuitive reasoning, algorithmic analysis, and reflection were selected. Through integrating them with the ideas given by the architects we developed the concepts of learning activities in the course.

Data on the effect of the environment were collected by:

- Design project portfolios. Portfolio evaluation included content analysis of the project activities and assessment of design solutions and mathematics applications. The design assessment criteria were based on the existing practice of studio evaluation and referred to the three following aspects: concept, planning/detailing, and representation/ expression. The mathematics assessment criteria were: perception of mathematical problems, solving applied problems, precision in drawing geometrical objects, accuracy of calculations and parametric solutions. Frequencies and correlations of grades in design vs. mathematics evaluation grades were examined.

- Attitude questionnaire and interview. The post-course questionnaire asked students to list the mathematical concepts studied in the course, give their opinion about its importance, and evaluate the learning subjects and methods. The in-depth interview with one of the students in the end of the course focused on his experience of applying mathematics in design before and in the course.

Analysis of results

Interviews and literature.

Design stages. In the first stage of the study, two experienced practicing architects were interviewed. They described their considerations in professional activities and relevant mathematical concepts, throughout the design stages: concept design, data collection and analysis, design alternatives development, design criteria formulation, design solution selection, models and drawings producing and presentation, solution examining and revising. This sequence of stages is similar to that proposed in [Hanaor 1998].

Design criteria and related mathematical concepts. The criteria and mathematical concepts identified in the study include the following: aesthetics, geometrical form, space division, proportions, functionality, culture, environment, symbolism, climate, geology, topography, construction rules and processes, people flow, energy, materials, stability, durability, building limitations, efficiency, modularity and accuracy.

Geometrical forms in architectural design. As noted by the architects, when developing structural forms they rely on the above mentioned criteria and answer questions such as: "How does the structure work? How does it fit in the entire project and in the environment? Is it a heavy/concrete or light/steel construction? Does it fit the concept?" They apply mathematical concepts such as reflection, symmetry, function, fractals, topological features, chaos, proportion, equality, identity, scale, algebraic surfaces, surface area, dimensions, volume and polyhedra. The literature sources mentioned in the geometrical forms section of the paper present numerous applications of these mathematical concepts in architecture design.

Mathematical aspects and course contents. The architects recommended teaching the mathematics concepts through architecture design projects that deal with curved surfaces, transformations, large structures and spans in airports, stadiums, etc. They emphasized the importance of mathematical methods in obtaining accurate design solutions. They suggested introducing the mathematics activities in the architecture design studio and then focusing on inquiry and learning discovery, design experiments and critical discussions. The architects proposed learning about geometrical objects in the order of their geometrical complexity – from point to line, to plane, to surface, and finally to volume. In addition to these recommendations, relevant teaching methods using context, visualization, heuristic and intuitive reasoning, algorithmic analysis and reflection, were selected from educational literature.

Grounded in the characteristics extracted from interviews and literature, we developed the concepts of learning activities in the course, defined a hierarchy of architectural and developed the MAAD course curriculum.

Design Project Portfolios

Each of the students in the MAAD course performed three projects and reported them in project portfolios. In this section we will consider the curved surfaces project, which is described in the "Case Study Framework" section of this paper.

Curved sufaces design project – grades. Our focus in evaluating student projects was on the correlation between students' achievements in design and in mathematics. Tables 3A and 3B present results of project assessment – average grades and standard devisations (S.D.) – following the design and mathematics criteria mentioned in the tables.

	Constructive efficiency	Aesthetics aspect	Architectural functionality	Program quality	Design grade
Average	76.9	85.6	73.5	87.5	78.9
S.D.	13.6	14.3	11.0	12.1	9.1

Table 3A. Curved surfaces design grades (%)

	Gas station dimensions calculation	Parametric analysis of surfaces	Roof model calculation	Building the physical model	Precision of model and calculation	Mathematical model analysis	Geometrical complexity of solutions	Math grade
Average	76.9	68.8	81.3	76.3	81.9	81.9	85.0	78.9
S.D.	13.6	13.8	10.4	9.9	13.8	12.5	11.1	6.0

Table 3B. Mathematics grades (%)

Tables 3A and 3B reveal the following features:

1. In the design assessment the students achieved high average grades for the program quality (87.5) and aesthetics (85.6) – the skills that they already acquired in the architecture courses. The grades for functionality (73.5) and efficiency (76.9) are lower because their insufficient experience in mechanics and advanced design.

2. The highest average grade among the mathematics evaluation factors was achieved in the geometrical complexity of solutions (85.0%). In spite of the relatively low credit given for this factor in the project assignment (8%), the students developed complex solutions since they were internally motivated to acquire experience in designing complex shapes. The average grades for the model related criteria (81-82) are higher than that for drawing calculations criteria (69-77). The possible reason is the educational advantage of creating real physical models which is emphasized in the educational literature [Oxman, 1999].

3. Close correlation between the individual design and mathematics grades was found, $\rho = 0.698$. This result indicates the tight integration of design and mathematical aspects of the course project.

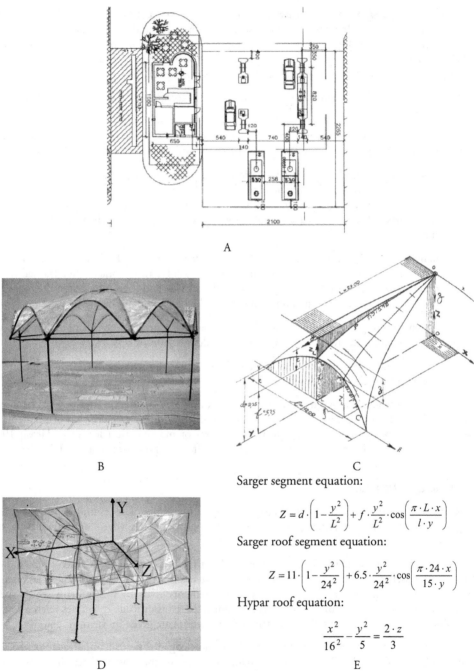

Sarger segment equation:

$$Z = d \cdot \left(1 - \frac{y^2}{L^2}\right) + f \cdot \frac{y^2}{L^2} \cdot \cos\left(\frac{\pi \cdot L \cdot x}{l \cdot y}\right)$$

Sarger roof segment equation:

$$Z = 11 \cdot \left(1 - \frac{y^2}{24^2}\right) + 6.5 \cdot \frac{y^2}{24^2} \cdot \cos\left(\frac{\pi \cdot 24 \cdot x}{15 \cdot y}\right)$$

Hypar roof equation:

$$\frac{x^2}{16^2} - \frac{y^2}{5} = \frac{2 \cdot z}{3}$$

Figure 1. A. Gas station plan 0:00; B. Sarger roof physical model for the pumps area; C. Sarger segment; D. Hypar roof physical model for the office area; E. Equations

Curved surfaces design project – example. Each of the project portfolios included the concept explanation, a module drawing, the module allocation plan, and a description of design and mathematical solutions. Here we present an example of project work produced by two students. The example includes a gas station plan drawing, performed by the students together, two roof models which they created individually, and equations (see fig. 1).

Our comments with regard to the design process description given by the students are as follows:

–   The student applied knowledge of algebraic surfaces in order to design roof surfaces of desired curve shape and implement the precisely in the physical models.

–   The drawings, models and mathematical descriptions given in the project report indicate that the students understand how to define and analyze surfaces, and their sections by coordinates. The students also demonstrate an ability to synthesize analytically surfaces which have desired properties and answer given constrains, such as Sarger and hypar surfaces.

–   When designing the gas station roofs, the students revealed the limitations of the CAD software in precise drawing of curved surfaces. To solve this problem they learned how to draw mathematical surfaces using the MatLab software and used it to design the Sarger and hypar surfaces which are then implemented in the physical models.

–   When building the physical models, the students acquired experience of dealing with efficiency and construction stability factors in design solutions.

–   The solutions presented in fig. 1 were designed by the students through iterations and selection of alternative variants.

The comments given above refer mainly to the design process. In the study we also examined the mathematical learning process as it was reported in the student's description of mathematical solutions. This analysis revealed that principal features of mathematical learning [Tirosh 1999], such as algorithmic, formal, intuitive, logical, and affective processes, appeared in the curved surfaces design context.

Attitude Questionnaire

The post-course questionnaire included four open questions. The examination of students' responses is based on categorizing the answers through context analysis and calculating their frequencies.

The first question evaluated students' opinion about the importance of the three course subjects (design of tessellations, curved surfaces, and the intersections of solids) for their architecture studies. The first column of Table 4 cites typical statements given by the students which specify aspects of the course contribution. The second column presents attitude categories related to these contributions. The frequencies of the contribution categories are given in the third column.

The absolute majority of the students (93%) acknowledged the great importance of the three course subjects. They identified the seven main categories of the course contribution which are presented in Table 4. The frequencies shown in the table indicate the percentage of students who mentioned certain categories as especially important for their cases. The majority of students emphasized the course contribution to deeper thinking on the subject of tessellation (category 3),

---

understanding mathematical concepts applied in architecture (category 1), designing geometrical forms (category 4) and connecting mathematics and architecture (category 5).

The second question of the attitude questionnaire asked every student to list mathematics concepts that he/she learned in the course. The answers are given in Table 5.

Attitude statements (aspects of the course contribution)	Categories of contributions	Frequency of categories (%)
1. *The course contributed to deeper understanding of golden section, Sarger surface, hypar and other concepts that were introduced in the Art and Architecture History courses.*	Understanding mathematical concepts applied in architecture	76
2. *After studying the course we are able to make a deeper analysis of structures with regard to additional aspects.*	Skills acquisition for analyzing architecture works	36
3. *Tessellation design is an important part of an architect's professional activities. This experience facilitated deeper thinking on the subject including module design, proportions, geometrical forms, symmetry, and harmony.*	Deeper thinking on the tessellation subject	84
4. *Mathematics is necessary for architectural design. How to make it precise? How to diversify geometrical forms? How to define structure contours without mathematics?*	Designing geometrical forms	64
5. *The unexpected discovery from the course was the connection between mathematics and harmony, aesthetics, and efficiency in various areas such as biology, music, anatomy, art and architecture.*	Discovery of universal mathematics formulations for harmony, aesthetics, and efficiency	88
6. *The course contributed to the general background knowledge, the connections between mathematics and different subjects such as botany and music were very interesting.*	General background enhancement	44
7. In design we need less specific calculation and more structured thinking and systematic consideration. In this project I acquired the ability of step-by-step design.	Structured thinking and systematic consideration of the project	24

Table 4. Students evaluation of the course

Mathematics concepts	Frequency (%)
Proportions, sequences, logarithmic spirals, polygons, symmetry, harmonic division, algebraic surfaces and line intersections (polyhedral, cylindrical, spherical, elliptic, and conic)	90-100
Cartesian and polar coordinates, circles and arcs, exponential and logarithmic functions	70-89
Similarity of triangles, irrational numbers, geometrical dimensions	50-69
Fractals, trigonometric functions, derivatives	30-49
Parabolas, limits, radians, tangents, equations and inequalities, differentiability, vectors and matrices	Less than 30

Table 5. Mathematical concepts learned in the course

Table 5 shows that the students in the course were exposed to a variety of mathematics concepts learned in class or on a need-to-know basis.

The third question related to the impact of the course on students' attitudes toward mathematics. The answers are summarized in Table 6 (which is similar to Table 4).

Attitude statements (aspects of change)	Categories of attitude change	Frequency of categories (%)
1. *I always had fears of mathematics, also after the first year course even though it differed from the school subject. In the course, mathematics has become so friendly, relevant and interesting that I succeeded in applying it.*	Finding interest in mathematics and its relevance	68
2. *The atmosphere of projects competition in the course motivated me to apply diverse geometrical forms, and this caused me to study functions deeper than I expected from myself.*	Recognizing the challenge of mathematics application	36
3. *When studying Calculus I asked myself: why do I need it, as I will never use it. And to my great surprise, I opened my calculus note-book looking for formulas that could help me in the project.*	Self-directed mathematical learning	48

Table 6. Attitudes toward mathematics

The majority of the students (72%) noted that the course changed their attitude towards mathematics. Almost all these students (68%) affirmed that the course aroused their interest in mathematics and demonstrated its relevance to architecture. Some of the students recognized the challenge and even looked for new applications of mathematics in architecture by their own.

The fourth question asked students to evaluate instruction in the course. As found, 64% of the students think that the studio-based instruction increased their motivation to learn mathematics, for 84% it stirred their interest, curiosity and was a challenge. The studio method enhanced students' creativity (60%) and opened a skylight to mathematics (68%).

Conclusions

Our longitudinal study shows the positive change of students' ability to apply mathematics to architectural design as a result of integrating the mathematics and architecture design curricula. The study started from developing the first year calculus-with-applications course based on the Realistic Mathematics Education approach. The course follow-up revealed significant improvement of learning achievements in mathematics and attitudes towards the subject. However, reviewing graduate architectural design projects performed at the college revealed that students scarcely used mathematical tools acquired in the first year mathematics course. In their design solutions the students avoided applying complex forms and surfaces in their design solutions.

In order to encourage students to use mathematics in design projects, we continued the integration of mathematics and architecture education by developing and evaluating the second-year MAAD course based on the MSS approach. This course offers mathematical learning as part of hands-on practice in an architecture design studio. It deals with three aspects of complexity in geometrical objects for architectural design: (1) arranging regular shapes to cover the plane (tessellations); (2) bending bars and flat plates to form curved lines and surfaces (deformations); (3) integrating and subdividing space by solids (constructions).

This paper focuses on the second aspect of geometrical complexity and considers the process of project-based learning of curved surfaces. The 20-hour curved surfaces design course consisted of three sections: mathematical concepts and methods with connections to architecture, practice in mathematical analysis of curved surfaces for architectural design, and a design project.

In the 2002-03 course follow-up study we used qualitative (ethnographic) methods, which observed learning behaviour within the context of the design studio using observations and interviews, attitude questionnaire, and project portfolios.

Our observations showed that the students approached the project experience with curiosity and motivation, and interest in deepening studies in mathematical subjects and their use. Assessment of students' activities in the projects indicated that the majority of them refreshed and practically applied their background mathematical knowledge. They also learned on a need-to-know basis and applied algebraic surfaces such as ellipsoid, elliptic hyperboloid, hyperbolic paraboloid, Sarger segment, etc. The correlation between design and mathematics grades showed the tight integration of the two subjects in the projects.

Analysis of attitudes questionnaires revealed students' high positive evaluation of the course. The majority of the students noted that the course aroused their interest in mathematics and demonstrated its relevance to architecture. The studio method encouraged students' creativity. Some of the students recognized the challenge and even looked for new applications of mathematics in architecture by their own.

Acknowledgment

The study was supported by the *Samuel Neaman Institute for Advanced Studies in Science and Technology.*

References

ALSINA, K. and J. GOMES-SERRANO. 2002. Gaudian Geometry. Pp. 26-45 in *Gaudi. Exploring Form: Space, Geometry, Structure and Construction,* D. Giralt-Miracle ed. Barcelona: Lunwerg Editores.

ALSINA, K. 2002a. Conoids. Pp. 88-95 in *Gaudi. Exploring Form: Space, Geometry, Structure and Construction,* D. Giralt-Miracle ed. Barcelona: Lunwerg Editores.

———. 2002b. Geometrical Assemblies. Pp. 118-125 in *Gaudi. Exploring Form: Space, Geometry, Structure and Construction,* D. Giralt-Miracle ed. Barcelona: Lunwerg Editores.

BOLES, M. and R. NEWMAN. 1990. *Universal Patterns. The Golden Relationship: Art, Math & Nature.* Massachusetts: Pythagorean Press.

Burt, M. 1996. *The Periodic Table of The Polyhedral Universe.* Haifa: Technion – Israel Institute of Technology.

CONSIGLIERI L. and V. CONSIGLIERI. 2003. A proposed two-semester program for mathematics in the architecture curriculum. *Nexus Network Journal* **5**, 1: 127-134.
http://www.nexusjournal.com/Didactics_v5n1-Consiglieri.html

ELAM, K. 2001. *Geometry of Design.* New York: Princeton Architectural Press.

FREDERICKSON, G. 1997. *Dissections: Plane and Fancy.* Cambridge: Cambridge University Press.

GILBERT, H. 1999. Architect – engineer relationships: overlappings and interactions, *Architectural Science Review* **42**: 107-110.

HANAOR, A. 1998. *Principles of Structures.* Oxford: Blackwell Science.

HOWSON, A., J. Kahane, P. Lauginie, and E. Turckheim. 1988. On the Teaching of Mathematics as a Service Subject. Pp. 1-19 in *Mathematics as A Service Subject,* A. Howson et al., eds. Cambridge: Cambridge University Press.

GRAVEMEIJER, K. and M DOORMEN. 1999. Context problems in realistic mathematics education: a calculus course as an example. *Educational Studies in Mathematics* **39**: 112-129.

KAPPRAFF, J. 1991. *Connections. The Geometric Bridge between Art and Science.* New York: McGraw-Hill.

KUHN, S. 2001. Learning from the Architecture Studio: Implications for Project-Based Pedagogy. *International Journal of Engineering Education* **17**, 4-5: 349-352.

LEWIS, R. K. 1998. *Architect? A Candid Guide to the Profession.* Cambridge MA: MIT Press.

LUHUR, S. 1999. Math, logic, and symmetry: construction, architecture and mathematics. http://www.sckans.edu/math/paper3.html (accessed 19 December 2004).

OXMAN, R. 1999. Educating the Designerly Thinker. *Design Studies* **20**: 105-122.

PEDEMONTE, O. 2001. Mathematics for architecture: some European experiences, *Nexus Network Journal* 3, 1: 129-135. http://www.nexusjournal.com/Didactics-Pedemonte-en.html (accessed 19 December 2004).

POLLAK, H. 1988. Mathematics as a Service Subject – Why? Pp. 28-34 in *Mathematics as A Service Subject* A. Howson et al., eds. Cambridge: Cambridge University Press.

RANUCCI, E. 1974. Master of Tessellations: M.C. Escher, 1898-1972. *Mathematics Teacher* **4**: 299-306.

SALINGAROS, N. 1999. Architecture, Patterns, and Mathematics. *Nexus Network Journal* **1**: 75-85. www.math.utsa.edu/s.../ArchMath..html.

———. 2000. Complexity and Urban Coherence. *Journal of Urban Design* **5**: 291-316.

SCHOEN, D. A. 1988. The Architectural Studio as an Examplar of Education for Reflection-in-Action. *Journal of Architectural Education* **38**, 1: 2-9.

TIROSH, D. 1999. Intuitive rules: A way to Explain and Predict Students' Reasoning. *Educational Studies in Mathematics* **38**, 1: 51-66.

UNWIN, S. 1997. *Analyzing Architecture*. London: Routledge.

VERNER, I. and S. Maor. 2003. The Effect of Integrating Design Problems on Learning Mathematics in an Architecture College. *Nexus Network Journal* **5**, 2: 111-115. http://wwwnexusjournal.com/Didactics-VerMao.html.

WILLIAMS, K. 1998. Preface. Pp. 7-8 in *Nexus II: Relationships between Architecture and Mathematics*, K. Williams, ed. Fucecchio (Florence): Edizioni dell'Erba. http://www.nexusjournal.com/conferences/N1998-Williams.html.

About the authors

Igor M. Verner received his M.S. degree in Mathematics from the Urals State University (1975) and a Ph.D. in computer-aided design systems in manufacturing from the Urals Polytechnical Institute (1981), Sverdlovsk, Russia. He is a certified teacher of mathematics and technology in Israel and a Senior Lecturer at the Department of Education in Technology and Science, Technion – Israel Institute of Technology. His research interests include experiential learning and E-learning, computational geometry, robotics, design, spatial vision development, mathematics in engineering and architecture education. He supervised the M.S. and Ph.D. studies of Sarah Maor, completed at the Technion in 2000 and 2005, in which architecture college courses "Calculus with applications" and "Mathematical Aspects of Architectural Design" were developed, implemented, and evaluated. Results of the studies were described in the papers published in the International Journal of Mathematics Education in Science and Technology (2001 and 2005), and in the NNJ (2003). The studies were presented at the International Conference on Applications and Modeling in Mathematics Education in London (2005). In Fall 2005 Dr. Verner worked as a visiting professor at the Tufts University and in January 2006 as a visiting scholar at the University of California, Berkeley. He gave talks on mathematics in architecture education at the Tufts MSTE Program seminar and at the Harvard Mathematics Education Seminar Series.

Sarah Maor received her Ph.D. in 2005 from the Department of Education in Technology and Science, Technion and is a Lecturer at Hadassa-Wizo College of Design, Haifa, Israel

Jean Brangé

École Spéciale d'Architecture
254 B d Raspail
75014 Paris FRANCE
jean_brange@esa-paris.fr

An Introduction to Algorithms and Numerical Methods Using Common Software

Jean Brangé describes an approach using Photoshop, VRML and C4D to allow architecture students to manipulate geometric and algebraic formulas, recursion, random functions, statistics, splines, the fourth dimension and other complex mathematical concepts.

Introduction

This paper, which was first presented at the Nexus 2002 conference in Óbidos, Portugal, will explain how to introduce algorithms and some complex mathematical concepts to architecture students without getting into practical math, but by using practical software functions which implement these concepts, allowing the students to manipulate math with the tools they are used to, and without the need to buy an expensive "Mathematica"-level software. In about 80% of the cases the students have no special interest in mathematics or in programming and one of the challenges is to figure out how to get them to work on these topics without boring them. I do not intend to prove or show that using numerical methods through algorithms is THE way to produce architecture, neither that c-space needs to be architectural, this would need another more complex paper; the bibliography contains some pointers to these issues.

I have developed this approach in my class in Paris, at the Ecole Spéciale d'Architecture, using Photoshop, VRML and C4D to get the students manipulate geometric and algebraic formulas, recursion, random functions, statistics, splines, fourth dimension, etc. The purpose of the class is to familiarize the students with mathematical and geometrical concepts outside the acknowledged "Euclidian 3 dimensions" they have been educated in and make them aware of the power of the mathematical, geometrical or analytical models, together with the necessity of defining a good "model" as a basis of a good architecture project. My definition of model is not the 3D model nor the wooden model but the logical or conceptual model. The goal is to get the students to define sets of rules that can generate a shape. The design is in the process and the definitions of the rules. This course is not an architecture course, so even if I try to connect the students' work to their actual studio projects, I do not spend time in the validation of the original constraints or intuition choices; we concentrate rather on the validity of the process.

Context

I started this course in the second semester of 1999. For some time I have been involved in redefining the general pedagogy of the school regarding the role of computer graphics (CG). This has led to a redefinition of the Computer Graphics courses in the school (France is quite behind in this topic compared to the US), with a set of introductory and instrumental Computer Graphics courses in the first and second years so that the students are all computer literate by the end of the second year. This includes "Office"-like, PSD and Internet software in first year, and 2D drafting and 3D modeling in second year. This would be the answer to "how" do we use computers. Starting in the fourth year we have set up a seminar on simulations, which is the "why" we do use computers.

Nexus Network Journal 8 (2006) 107-111
1590-5896/06/010107-5 DOI 10.1007/s00004-006-0005-y

Lab

The computer lab is divided into three sections. The main part is the classroom: divided in two parts face-to-face and equipped with twenty seats – half Macs, half Pcs – with two video projectors and two teaching seats. On each side of the classroom we find two rooms: one for students working on their final thesis or a long-term work, and one used as a self-service room. Related to the computer lab we have an Intranet based on the First-Class workgroup solution, which gives us collaborative work places for any course, web publishing facilities, and a general bulletin-board system. The workgroup server is also available through the web and is replicated on an external server, so we can access the bulletin board from anywhere. Dial-up access to the Intranet is also available through Apple remote Access protocol. The software used in this hardware and network solution has been chosen for its pedagogical efficiency. We did not follow the standard market shares but the ease of use, understanding and appropriation by the user. The set of software we base our courses on includes Photoshop, GoLive, Word, Excel, IMovie, VectorWorks, Cinema4D and Archicad. Other softwares are available and installed – AutoCAD, Architectural Desktop, 3Dmax, Artlantis, Radiance, Illustrator, etc. – but these are not taught within a specific course, and are more used as self-training solutions.

Reform

One of the goals of the reform we are setting up is the strict definition of the vertical coherence of the disciplines in the curriculum and their inter-related horizontal connections. Horizontal connections are available through the fourth-year seminars, which allow outsiders to participate. For instance, the "Figuration" seminar has produced conferences on the evolution of drawing techniques up to Computer Graphics systems.

Example 1. Automation as "fast to execute repetitive task" and "batch" is one of the main characteristics of the computers. Nowadays young people do not feel the power of modern computers since they are already accustomed to it, so they are not surprised when a renderer computes a picture in a few minutes. However, they are shocked when they find out that some settings can blow the rendering up to a few hours. By asking computers to produce quickly repetitive tasks they rediscover why programming is really useful. Photoshop is the best tool I have found to explain this concept to architecture students. Photoshop's macros and batch utilities are a simple and yet powerful development system to work on pictures, which can be used for picture manipulation but also for image analysis.

Photoshop also provides a large set of predefined functions, which can be called up by commands in the "filters" and image "adjust" menus. The Photoshop macro program also makes it possible to include some user interaction by input dialogs. This allows students to understand the sequential programming style and the use of variables.

The main exercise consists in getting a large set of pictures, acquired by any means available to the students – screenshots, digital video, cellular automata software, etc. – and to apply a repetitive process to these pictures that will lead to the definition of a 3D shape, built from the translation in 3D space of the isovalue curves of the pictures analyzed.

The first part of the exercise is to build a Photoshop script, which will transform the raw picture into a set of 2D curves. The script is a simple macro recording of a succession of steps defining image manipulation functions, which can include, brightness and contrast, resolution, selections, focus, color manipulation, noise filtering, black/white levels, thresholds, etc. The list is not limitative.

From the filtered image, we can select a range of values to extract a 2D curve or surface. This curve is then sent to a 3D Software for 3D generation of a surface or a volume. Once the script is defined it can be applied to a large set of original pictures to produce a new set of 2D curves or surfaces. The process is set up. The script can be saved and transmitted, modified, included in more complex constructions. Most of the script steps can include stop points and user interaction for fine tuning or parameter input. The result is the automated production of sections which can define a 3D mesh from a bunch of raw data. Two examples of this exercise are joined in the graphical section.

This exercise implements the automation, the sequence, the function calls, the user interface including user input and data exchange, but we still miss branching, looping and Object Oriented concepts. VRML will be used for that.

Example 2. VRML as derived from the Silicon Graphics OpenInventor Format is a toolkit for developing interactive graphics application and includes some algorithmic features such as looping and conditional branching through specific VRML Nodes, such as the Switch, Level Of Detail, and Sensors. Further, the VRML Script Node enables the student to create specific functions that share parameters with the other VRML nodes.

Thanks to these Algorithmic tools we are now able to deal with some more complex notions as recursion and fractals while using VRML. VRML includes scripting among the other fundamental modeling elements, such as "box", "cylinder", or "texture", etc. Using scripts we have access to nearly unlimited object-oriented programming functionalities inside our models, but this tends to be quite difficult to transmit, since we need to get into real algorithm codes and language syntaxes. We can still approach some basic algorithm concepts, without writing complex codes. Some VRML nodes implement basic algorithm concepts as branching, switching, functions and loops.

Simple loops can be implemented by using a time sensor, that is, loops based on a running clock, whose increment and step can be set at our convenience.

Other sensors like proximity sensor or visibility sensor, as well as the "Level Of Detail" node are based on a "if/then/else" sequence,

```
if I see the object
then do that
else do that

if I am within a certain distance of an object
then do that
else do that
```

The switch node implement is a "switch/case" concept :

```
Case x:
    1: show object 1
    2: show object 2
    3: show object 3
```

By using these specific nodes together, students can achieve complex process definitions, where the model responds dynamically to the user interaction. The most basic interaction is the navigation of the user in the model.

Observations

Computation power. The data analysis forces the student to work quickly on a large amount of files. Since we are able to produce rather large models and since we use them in real-time interactive software, we quickly reach the power limit of our available hardware. Both exercises point out the power (and the weakness) of the computers.

Trans-disciplinarity. Some students use the results of this course in some other classes, not only in the architecture studio. Reciprocally they apply in this course ideas and concepts borrowed from other disciplines.

Curiosity on these new territories for architects. Some students choose to pursue the work while preparing their final thesis a project where design process is really an issue, and usually this involves other types of "built" environment. The unbuilt, ruins, light, networks, media, cyber space, etc., may be used as new architecture materials.

Limits. The basic limits that we encounter are algorithm, math, statistics and time.

We do not have enough time to have a real algorithmic course of even only 12 hrs. The data analysis we simulate are only useful as a simulation of the process. The real process is not done on real data, as we do not have access to the real data. On the math level, we are limited to simple arithmetic functions when we have to code them ourselves, and we cannot go into the detail of the more complex ready-made functions available in the software we use.

Another limit is the amount of prior architectural work needed to start the process. The course is only an elective course. Last semester, the topics of the exercises were individually connected to the students' projects in their architecture studios and the results were less interesting in the CG course than in the Architecture Studio. The students explored a lot but were not able to produce a specific part of their studio work using the CG course as requested. The best results were obtained by students who had quickly jumped into the "processes" to set-up a starting point for their architectural project. This has led to interesting development in the studio, but we were not able to establish a bidirectional link between the studio and the lab. The studio's "time" caused a huge slowdown in the middle of the semester (time was needed to refine the projects, and at that point, the dexterity of the students with their new "dynamic" C4D software was not highly enough developed). This should change in the next couple of years since all the students will be provided with a good knowledge of 2D, 3D and animation, acquired in the first two years. For us, the best results were achieved when the CG course had its own "architectural" exercise disconnected from the current project studio.

Conclusions

Are we in presence of a new tool used with the same methods and concepts, or is it the same tool while the methods are new? Can we break the ancient rules? Are we setting up new rules? Can we set up "rule breaking" as a rule?

Efficiency. By using our chosen tools we can bring the students to a faster understanding of the computer graphics concepts. They will therefore spend more time on "how" and "why" to use these tools, rather than on "where is the function I need?", or "is there a function to do what I want?". We have observed that some students are switching from 3DSmax to Cinema4D, although they have worked with 3DS for a few years. They discover that a simpler but better designed tool is more efficient. All the students admit that the C4D interface is much less time consuming and gives more direct access to the tools and functions available. The result is that in a few sessions students with no previous 3D modeling knowledge become able to model their Architecture Studio project and to build a final full polygonal, nurbs, Boolean, texturized, and lighted model.

Process as an element of design. The basic idea of these exercises is to formulate in non-graphical language how a situation can be approached and how we can interact with it. This defines the process, and this can set the roots of the design intuition.

Measure. We might use this process as a system to try to find in the object the definition of the object itself. If a human being can be considered as a sample of a living species defined by its DNA sequence, the DNA sequence can be imagined as a computational object with the data, the program and the processor all included in one entity. The fact that the DNA sequence is made out of the simplest alphabet we could imagine does not give us the clues to the infinite combinations that are available within these sequences. But still the DNA defines itself. In the same way, by being able to define a proportion system, an organization scheme, a procedural pattern in terms of process, and letting this process model a shape, the shape will contain in some case the basic structure of its processing development in a human readable form.

Bibliography

Publications:
Architects in Cyberspace. 1995. *Architectural Design* 65, 11/12.
BENEDIKT, M. 1993. *Cityspace, Cyberspace,and the Spatiology of information.* Austin: University of Texas Press.
MITCHELL, W. 1994. *City of Bits.* Cambridge MA: MIT Press.
MOUSSA, Arnaud, et al. 1997. Meta-architecture en France. *kubos*.
    http://www.kubos.org/vrml/metarchi.html

Websites:
Centrifuge. http://www.construct.org/motet
Marcos Novak. http://www.centrifuge.org/marcos
Objectile. http://www.objectile.com/
Assymptote. http://www.asymptote.net

About the author

Jean Brangé teaches computer-aided architectural design at the Ecole Spéciale d'Architecture of Paris. His own research focus on the applications of computer science in architectural drawing and design. He collaborated in the development of some specific softwares (Radiance, Octree and Arris, among others). He is also involved in the development of internet technologies for professional purposes and is a specialized consultant in PLM for the French association of the world-wide-web users ("Afnet").

Michela Rossi

Sede Scientifica Facoltà di
Ingegneria, Palazzina 9
P.co Area delle Scienze,
181\A
Parma – ITALY
michela.rossi@archiworld.it

Natural Architecture and Constructed Forms: Structure and Surfaces from Idea to Drawing

This work grew out of didactic experience in architecture classes at the universities of Florence and Parma. The comprehension of geometric schemes in regular organic objects formed the basis of teaching drawing and scientific representation, such as formal architectural synthesis. This exercise may offer also a valid starting point to help students approach mathematics, and help them to imagine and plan the increasingly complex surfaces of late contemporary architecture.

Drawing and architecture

Just as mathematics speaks through numbers, drawing is the compositional and essential language of architecture. This language is used to express concepts through simple planar or spatial entities, correlated to forms generated by geometric thought.

Because of this formal identity of fundamental entities a close relationship has developed between geometry and architecture. This relationship is evident in the articulation of built form, in which diverse applications of mathematical models are found, used to resolve problems of both statics and aesthetics.

The connection is particularly clear in drawing, in which references to geometry are used to explain and describe complex forms. In fact, drawing and geometry link architecture to mathematics and express the same concepts, since they allow the form to be seen in reference to the elements of point, lines and planes.

Fig. 1-3. Villard de Honnercourt: geometric grids in drawing construction of natural forms

This fact is evident in the teaching of architecture, where since the age of Vitruvius the two disciplines have shared a pre-eminent role without necessitating any particular modification of language – of common usage – between mathematics, or geometry, and design and architecture.

1590-5896/06/010112-11 DOI 10.1007/s00004-006-0007-9

The *Sketchbook* of Villard de Honnecourt[1] is significant in the use made of drawing during the Medieval, of outlines and simple geometric forms. Various Gothic designs confirm the role of geometry in governing the statics of the structure through the use of the simplest plane forms: the square and triangle, used respectively in the design of plan and elevation (figs. 1, 2, 3).

Thus geometry offers a means of learning drawing and understanding meaning, since both drawing and meaning appertain to architecture.

Since we draw things by means of their apparent outlines and the discontinuous lines between surfaces, it is important to study the nature of their intersections. The principle problem resides in the difficulty of verifying the spatial characteristics of the most complex forms, such as are found in architecture: a meaningful example is the intersection of curved surfaces in vaults and domes.

The scientific solution of representation derives from the reduction of three-dimensional forms to the plane by means of projective and descriptive geometry,[2] but in actual fact architects resorted to plane sections for the study of three-dimensional objects, even before these were codified. The solutions are rather simple for simple surfaces such as planes, cones and cylinders, and even the sphere, but become more complicated as the degree of mathematical complexity grows. Sometimes we must superimpose a grid on the surface in order to describe the form by means of a regular network of points, as series of plane sections that are orthogonal or radial with respect to each other (fig. 4).

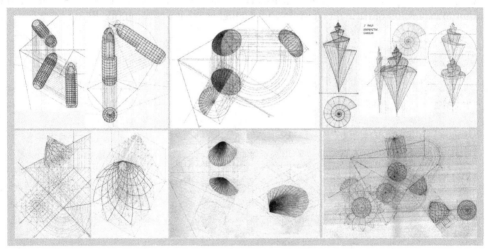

Fig. 4. Students' geometric exercises on forms and surfaces in natural architectures

In reality, architects have always resorted to the use to models, even wooden ones,[3] built according to the same constructive logic that they were intended to simulate, in a way that is analogous to how we create virtual models today, which are always based on compositions of geometric elements.

Virtual modelling no longer requires the use of Mongian projection to solve problems of representation and measurement, but in any case the construction of objects requires careful geometric control of solid forms and curved surfaces.

Once we know the rule (or rules) that govern the form and the relationships of its parts, it is no longer relevant if the instruments used for its description are pencil on paper, numbers and equations, or a virtual model. But since at the very beginning in drawing, in architecture, there exists only an idea that grows in our minds with the support of geometric references, we always need graphic notes for visualising the various stages of this development.

Usually architects prefer the pencil while engineers prefer numbers – this perhaps is the most obvious difference between the two – but in either case geometry is used to identify the new forms.

Many of the figures that illustrate the present paper are derived from experience in didactics conducted during courses of architecture at the universities of Florence and Parma. The study of regular natural forms was proposed as an exercise in Drawing and Descriptive Geometry. The didactic itinerary for understanding drawing and scientific representation was defined by the search for a geometric scheme of reference for a formal synthesis of the architecture of regular natural objects (figs. 5, 6, 7). This undertaking allowed the students to investigate how geometry characterises natural architecture, and furnished as well a stimulus for approaching mathematics and the study of the increasingly complex surfaces that characterise recent developments in architecture.

Beauty and geometry

Since antiquity man has been fascinated and awed by the beauty of the natural world, and have lingered over the regular conformations of crystals, living creatures (simple animals, plants and flowers) or their parts,[4] up to the Renaissance,[5] when the human body was considered the maximum expression of natural perfection and the highest divine creation (fig. 8).

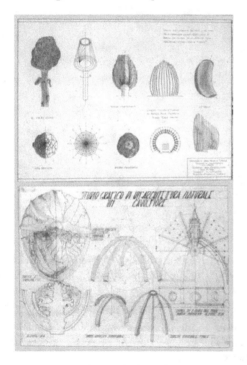
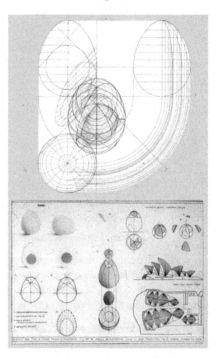

Figs. 5 and 6. Students' studies on geometric laws in architecture: natural and manmade constructions

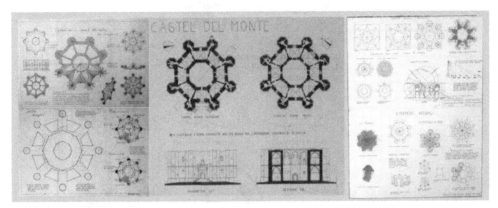

Fig. 7. Students' studies on geometric laws in architecture: natural and manmade constructions

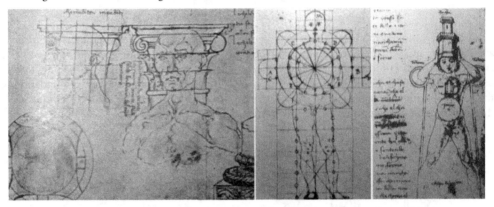

Fig. 8. Francesco di Giorgio Martini, human perfection in in architectural proportioning

According to tradition, in treatises from Vitruvius to those as recent as Gottfried Semper,[6] the origin of architecture was traced back to the imitation of nature, and for a very long time the formal inspiration for ornament and decoration were sought in the regular configurations[7] of natural examples. But above all architects found the solution to specific structural problems in nature. The architectural orders are perhaps the prime, as well as the most famous, but still only one example among many, since nature offers a great quantity of models that respond to structural, formal and aesthetic problems.[8] In spite of the passing centuries, what appears to be a game without any evident rational foundations is the repetitive reference to nature as a design model and a rule for aesthetic equilibrium, which led to the obsessive search for a rule of beauty based on geometry.

Since the principal motive for these similarities between structures depends on the force of gravity – which subjects all bodies both natural and built to the same laws of equilibrium – it is not surprising to find similar static schemes verified by analogous mathematical models. The model of the structural system can be as static (shells and ribs) as it is mechanical (the skeletons of vertebrates) and sometimes the natural architecture is much more complex than the manmade architecture, since buildings require neither movement nor velocity (fig. 9).[9]

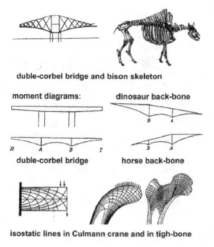

duble-corbel bridge and bison skeleton

moment diagrams: dinosaur back-bone

duble-corbel bridge   horse back-bone

isostatic lines in Culmann crane and in tigh-bone

Fig. 9. Analogies of statics in vertebrates and architectural structures

This observation demonstrates that the relationship that exists between natural and artificial architectures, in the common composition of parts according to rules governed by geometry and/or the growth of forms, underlines the concrete nature of the classic myth of the imitation of nature. Because of the various symmetries that exist in natural forms, the effective foundation of this presupposition is indeed geometry, capable of conferring harmony and equilibrium. It therefore becomes an important element of design and construction.

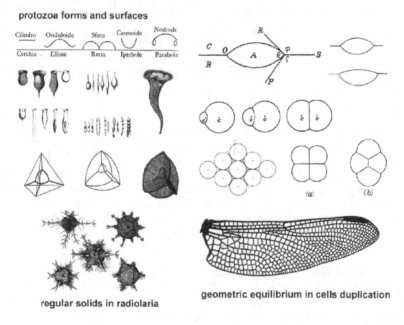

protozoa forms and surfaces

regular solids in radiolaria

geometric equilibrium in cells duplication

Fig. 10. Geometric surfaces and solid forms in cells: soap boubles, protozoas and biological tissues.

Effectively, mathematical models were developed to simulate reality by means of numbers, but geometry, which refers to form, is an concrete element of reality: everyone knows the logarithmic spiral of the Nautilus shell, the regularity of starfish, the perfection of the egg, and so on, but going beyond this, in protozoa are found living beings with the shape of all of the surfaces of Plateau, while radiolarian skeletons exhibit the forms of all five of the Platonic solids. It seems almost as though nature wanted to play with geometry (fig. 10).

Many centuries ago, Plato believed that all of reality could be traced back to two kinds of triangles, those found in the regular solids that symbolised the four fundamental elements. He didn't justify this, but he had to have been conditioned by the strong presence of geometry in natural objects. Much later, Kepler was fascinated by snow crystals and beehives. In more recent times science has explained the presence of constant angles in crystal structures through molecular chemistry, while regular organic forms are linked to the biological necessity cells, tissues and the growth of living organisms. Naturalists have explained that the Fibonacci series and the Golden Number effectively exist in plant and animal forms, because the mean ratio (Golden Section) satisfies the principle exigencies of growth, which is that of maintaining the same form and therefore the same harmonic equilibrium (as in the gnomon), important because a change in this equilibrium would require a search for a new vital equilibrium (fig. 11).[10]

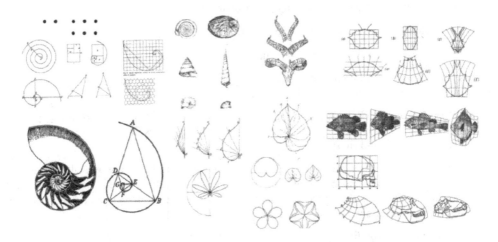

Fig. 11. Spirals and growth, maintenance and deformation of form

D'Arcy Thompson[11] explained that the main problem of natural phenomena is always linked to efficiency and the search for the minimum output of energy. He explains and illustrates the problems of minimal surfaces and saturation of space with geometric models that are resolved by means of the most elementary regular solids.

Just as similar forms, from the smallest to the very largest, appear in the architecture of nature, in a way that recalls fractals, self-similar elements of different scales characterise creations of human design, and regular geometries give evidence of an equilibrium between the static symmetry of closed forms and the dynamic movement in relation to asymmetric forces that belong to growth and life.

The substance is different, but the concept is not too distant from the Platonic idea ... and even Leon Battista Alberti was right in some way: in fact, if there is divine perfection in the human body and it exhibits the proportions of the Golden Number, this depends solely on the fact that growth must be regular, except in the cases of error or accident, and the Golden Ratio – and only the Golden Ratio – can guarantee this requirement.

As a consequence, the study of diagrams of its formal phenomena in nature – on which classic thought is based, and therefore the development of all of modern science – facilitates the resolution of many design projects, especially the research for Alberti's *concinnitas*, which is satisfied in the harmony of parts expressed by means of shape and number (geometry and arithmetic). This allows the satisfaction of the principle requisite of classic architecture, that is, the Vitruvian triad of *firmitas, utilitas, venustas*.

Formal models for design

The great variety of configurations in nature can be correlated to relatively few formal models based on different diagrams and symmetries, which make up the geometric basis of architectonic imitation. Both plane and spatial figures are always organised according to a simple diagram that can be traced back to three fundamental archetypes:

- Modular aggregation according to a regular grid;
- Radial division of a circular unit in polygons;
- Linear continuity of spirals as regular growth of forms.

These models exhibit different kinds of symmetry and logic in particular growth patterns, and each of them has specific geometrical rules.

Modular aggregation. Modular aggregation permits growth that is discontinuous and asymmetrical, according to the direction and the number of grid lines; it recalls histological tissue and can cover the plane and fill space, as well as expand linearly.

The symmetries according to which the base module is reproduced are several (translation, reflection, rotation...) and can combine with each other in very complicated ways, but they are always repetitive. The module is predetermined in relation to the fundamental grid diagram, but does not preclude a great diversity of solutions.

Growth is therefore conditioned by the module, and this modifies complex form of the whole, which is indeterminate and thus permits the greatest degree of liberty. We can observe these models in the drawings of surfaces, in relationship to ornament and wall structure, and in the modular aggregation of spaces in plan as well as in spatial composition. We find them in the shapes of surfaces, in the structural mechanics of constructions, and again as an ambiguous game between the drawing of the surface and the representation of space (figs. 12, 13, 14).

Radial division. Radial division exhibits a closed form and a repetitive symmetry with respect to the centre, which often has mirror symmetry, but not necessarily the same number of axes. Growth takes place only in an outward direction, thus it is discontinuous and remains concentric so that the form is predetermined. This model can have a radial grid or can be aggregated in relationship to other grids. The extensions in space of this model are identified with rotated closed solids, such as domes.

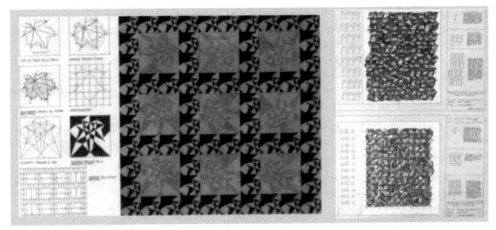

Fig. 12. Regular grids and surface organization: students' studies on regular patterns

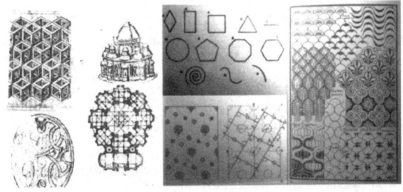

Fig. 13. Regular grids and surface organization: studies from Leonardo and Le Corbusier

Fig. 14. Regular grids and surface organization: Fritz Hoeger's ornamental brickwork

Linear continuity. Spirals derive from unidirectional linear growth that can be either planar or spatial, but which is usually refers to a continuity that tends to the infinite and to a particular rotational symmetry, which in logarithmic curves does not alter the proportions of the form. Thus growth is not discontinuous, and the form remains open. As a consequence of these conditions, man seems to have been particularly inspired by spirals since antiquity: spirals are used as special symbols of life and are manifest in art in different ways, while in architecture they become actual architectural elements.

Their spatial conformation generates complex surfaces that derive, however, from the regular motion of simple forms. Thus, since they are based on geometric rules, it becomes easy to draw, understand and construct the form.

These three models can be combined with each other in infinite combinations, each of which can be in its turn varied while maintaining homologous characteristics.

These geometrical schemes satisfy various requisites of construction, suggesting solutions both formal and structural, with numerous examples found at all scales in architecture, from urban and territorial planning to ornament and surface decoration.

The importance of the module

We all know the significance that the concept of the *module* has in architecture and its importance in measuring, which is precisely the relationship between unit and quantity. Since this concept is directly connected to the use of modular grids that govern composition and proportion, it can be said that the design project makes reference to the concept of measure, and that this takes place through geometry.

The module is the basis of architectural order, which is the first principle of structures, organised according to spatial grids with orthogonal directions. In architecture cubic and pyramidal grids are common; in the regular organisation of the plane as well as the articulation of surface there are many possible solutions, while spatial applications are more difficult, except in the imaginative fantasy as in the works of M.C. Escher (fig. 15).[12]

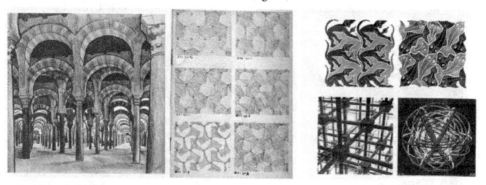

Fig. 15. Regular grids and surface organization: M. C. Escher's studies about surface and spatial grids

In spite of its being a closed form, solutions based on the radial scheme are numerous and diverse; in drawings of plane configurations, such as those of rose windows or pavement designs, the number of the divisions and concentric elements change. This plane scheme is often used in urban design and in the realisation of buildings with a centralised and hierarchical spatial layout, in which the formal articulation is reflected in adjunct minor spaces. In spatial forms this scheme generates domes that can be composed of surfaces conceived according to different design solutions, in relationship to the structural choices for the building: continuous shells, ribs or geodesic grids as in the work of Buckminster Fuller.

The spirals evokes continuity, and its shape, often conjoined to the Golden Ratio and charged with symbolic and aesthetic meaning, has always stimulated formal invention: we find it in designs for ornament and in architectural projects, where it satisfies the necessity of a continuous growth in space, such as in the Guggenheim Museum by Frank Lloyd Wright.

Double spiral grids are frequent in nature and are not unusual in architecture, as for example in the design of the surfaces such those in Michelangelo's Campidoglio in Rome or in the subdivisions of the domes of Guarini and Taut.

Fig. 16. Student study of irregular surfaces

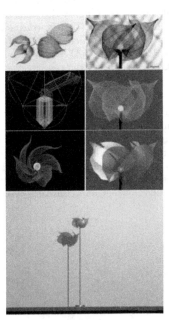

Fig. 17. Student design developing from natural form

Conclusion

Thus we see that geometric elements are the principal design tools used to make ideas of a project concrete. We find confirmation of this when we compare the marvellous variety of natural objects with architects' imitations of them, and at the same time we find stimulation for new projects.

In imitating organic forms and structures, architecture shows its age-old relationship with primary geometric elements: numeric sequences controlled by unambiguous laws that characterise natural construction as much as built objects. As the figures for this paper show, drawing is an expression of geometry for understanding and describing form. With regards to architecture geometry communicates more clearly than words, because it becomes the concrete aspect of our imagination.

Notes

1. Several sheets of his sketchbook show human faces or animal bodies built up on geometric grids; see Villard de Honnercourt, *Disegni*, Jaka Book, Milan, 1988.
2. Before Descriptive Geometry was codified by Gaspar Monge in the eighteenth century as a graphic application of Projective Geometry, orthogonal views were used in plan and elevation drawing, as shown in many medieval projects and some archeological objects.
3. Wooden models has lost their importance because plane omology allowed easy and more economical solutions to of spatial length problems using ortogonal projection.
4. Ian Stewart, *What shape is a snowflake?*, Weidenfeld & Nicolson, London, 2001.
5. Luca Pacioli, *De divina proportione*, Venice, 1494.
6. Gottfried Semper, *Der Stil*, München, 1860-63.
7. Ernst Gömbrich, *The sense of order,*1979.
8. Paolo Portoghesi, *Natura e Architettura*, Fabbri Editori, Milan, 1993.
9. D'Arcy W. Thomson, *On Growth and Form*, Abridged Edition, Cambridge, 1961.
10. Mario Livio, *The Golden Ratio*, Broadway, 2002.
11. Figs. 9-10-11 are from D'Arcy W. Thomson, *On Growth and Form*.
12. M.C. Escher, *His life and concrete graphic work*, Abradale, New York, 1982.

About the author

Michela Rossi Michela Rossi is a professor or architectural drawing and representation at the Architecture Department of the University of Parma, Italy. She earned her Ph.D. at the University of Palermo, discussing a dissertation that addressed the question of structure and ornament in historical and contemporary architecture. She teaches courses of descriptive geometry and architectural drawing, and pursues her own research on the characteristics of traditional architecture in Northern Italy, and on the transformations and historical evolution of landscape. She is author and co-author of several books, including monographs on Tuscan rural churches, on Renaissance *palazzi*, and on the network of rivers and canals in the Parma region.

Cornelie Leopold

Darstellende Geometrie und Perspektive
Fachbereich Architektur
Raum- und Umweltplanung,
Bauingenieurwesen
Technische Universität Kaiserslautern
Postfach 3049
D-67653 Kaiserslautern GERMANY
leopold@rhrk.uni-kl.de

Sound–Sights
An Interdisciplinary Project

An interdisciplinary project between geometry, architecture and music resulted in a concert and exhibition with sound installations. Professors and students of architecture and mathematics of the Technical University of Kaiserslautern worked with a professor and students of music composition of the Music Academy of Cologne in Germany. The theoretical and historical analyses of the relationships between geometry and music formed the basis for original creative works in interdisciplinary groups. Music was composed according to geometrical-architectural concepts and geometrical images, forms and processes were developed after musical ideas. Geometrical forms are combined with the music into a kinetic, visual and acoustic work of art. Through such interdisciplinary art projects it is possible to experience scientific coherence in a sensual way. The combination of geometry, architecture and music enables a visual and aural approach to formal thinking of sciences.

Introduction

Exploring the relationships between geometry, architecture and music formed the background of an interdisciplinary project with students and professors of architecture, mathematics and music composition. The aim of the project was a concert and exhibition with sound installations entitled "Sound–Sights. Seeing Music – Hearing Geometry" [Leopold 2003a; Leopold et al. 2004] held in the concert hall of the city of Kaiserslautern, Germany. The project was initiated by Andrea Edel, director of the cultural office of the city of Kaiserslautern. Theoretical and historical analyses of the relationships between geometry and music preceded the students' creative works in interdisciplinary groups. The groups were organized during a common workshop week where each student presented his/her first ideas. It was not possible in all cases to bring together interdisciplinary groups. Finally the various individual or group projects were arranged in the concert and exhibition program.

Fig. 1. Projects: "Space-Sound-Sphere", "Balanced Sound Sculpture" and "Abacus"

Music was composed according geometrical-architectural concepts, and geometrical images, forms and processes were developed according to musical ideas. Geometrical forms were combined with the music to a kinetic, visual and acoustic works of art. Multimedia computer technologies were used for the connection of image and sound in some projects. But the human being behind the creative process always remained visible. Combining the sense of seeing and the sense of hearing underlined the differences and similarities between audible and visual perception.

With interdisciplinary art projects such as this, it is possible to experience and mediate scientific coherence in a sensual way. The combination of geometry, architecture and music enables a visual and aural approach to formal and structural thinking in art and sciences.

Theoretical and historical background

The idea to look for the relationships between geometry, architecture and music goes back to the ancient understanding of the sciences where the seven *artes liberales* were grouped in the *trivium* (grammar, rhetoric, logic) and the *quadrivium* (arithmetic, music, geometry and astronomy). The study of harmony is a common element of the quadrivium. Architecture was seen as a mechanical art, in which harmony and proportion were applied to the principles of creating a building. Geometry and music were developed on the basis of the Pythagorean concepts. Especially in the European development of the sound systems, mathematics and music were closely linked. The development of the musical scales and the geometry of elaborating motifs shows the relationship between mathematics and music.

Pythagorean Harmony. For the Pythagoreans, music, arithmetic, geometry and astronomy were very closely related. "Harmony" had an extensive meaning and was related to philosophy, arts and sciences. It was understood as regularity, order and regular arrangement of many parts. The word "harmony" is derived from Greek *harmos*, which means combination, adaptation, connecting different or opposite things to a ordered whole. "Harmonia" is also a mythological person, the daughter of Ares, god of war, and Aphrodite, queen of love and beauty. She therefore represents the union of two opposites.

Fig. 2. Pythagoras with the tetraktys, from Raffael's *School of Athens* (Stanza della Segnatura, Rome) and Pythagoras with experimenting with sound [Hay 1968, 150]

According to Pythagoras, harmony and all things and principles of being can be expressed by integers and mathematical regularities. It is said that Pythagoras overheard the sounds and harmony of blacksmiths' hammers, leading to experiments with the division of a string. He found that musical intervals are achieved by the division of a string as well as the relations between the number of sound oscillations. The integers 1, 2 , 3 and 4 form the "tetraktys," which is the basis of all harmonic proportions. The sound experiments were developed by means of the monochord, a simple instrument with one string tightened over a resonance box. The proportion 1:2 is the octave, the proportion 2:3 stands for the fifth and 3:4 for the fourth. Pythagoras transferred these proportions to astronomy as well. He believed that the stars stand also in these proportions and they produce divine music, the sphere sounds. Later on, in the Renaissance, the tetraktys was expanded by Zarlino. The relationships between architecture and music were very close in the Renaissance. Alberti wrote that the rules for harmonic proportions in architecture should be borrowed from musicians [Naredi-Rainer, 1982; Wittkower 1971].

Inspired by the Pythagorean ideas and combined with new technologies, the projects "Abacus" as well as "Point and Line" were created by our students.

Transformational Geometry.   Another background for the creative work was found in investigations about applications of transformational geometry in music and the geometrical analyses of musical motifs. Transforming musical motifs according some rules is a method often used in musical composition. If we illustrate the transformations by pitch-time diagrams, we can interpret the transformations as geometrical transformations: transposing as translating in direction of a second axis, retrograde as a reflection at the vertical axis, inversion as a reflection at the horizontal axis and the retrograde inversion as double reflections, which can also be interpreted as point reflection [Christmann 2003; Leonhardt and Willenbacher 2003]. In this way concepts of symmetry are applicable to musical composition. Through this geometrical interpretation of musical motifs we perceive relations between music and patterns. Motifs with translations are for example interpretable as frieze ornaments. Fig. 3 shows such an example.

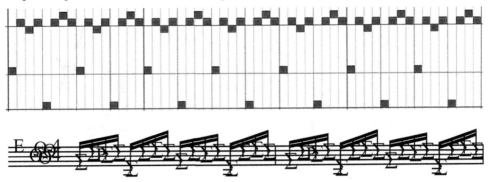

Fig. 3. Frieze ornament by translation, after B. Smetana, *Moldau*, bar 1+2, left hand, piano arrangement

The music of Bach especially lends itself to interpretation by transformational geometry. There exist many experiments through history of art to express the music of Bach visually.

Looking for geometrical structures in music by translating it into visual elements or by applying musical structures to visual material was another inspiring source in the students' projects; one example is "Ball," described below.

Formalisation of Aesthetics.   Symmetry concepts have always played an important role in aesthetics. By interpreting symmetry with the help of transformational geometry we have already taken a step towards the formalisation of aesthetics. Reflections about proportions and developing scales of proportions with the golden section, such as the "Modulor" of Le Corbusier, are other examples of developing formal systems in aesthetics. Architect and composer Iannis Xenakis (1922 –2001) worked to find mathematical solutions for artistic problems.

Xenakis worked in the atelier of Le Corbusier in Paris from 1948 to 1959. He applied the "Modulor" to architecture and to music. The rhythmical structure of his *Metastasis* was composed with increasing and decreasing density according the "Modulor"; in like manner the facade of monastery "La Tourette" was an application of the "Modulor" in the *pans de vers ondulatoires* [Leopold 2003b].

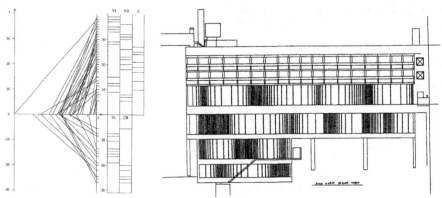

Fig. 4. Scheme of "Ondulatoires" in Xenakis's *Metastasis*; the west facade of "La Tourette"

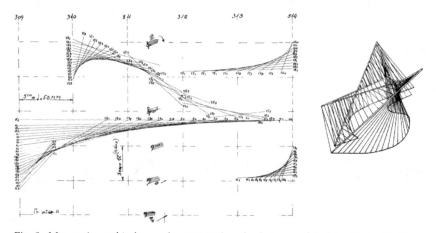

Fig. 5. *Metastasis*, graphical score, bar 309-314, and a design model of the Philips Pavilion.

Xenakis used geometrical abstractions as ordering principles in music and architecture. For example, ruled surfaces were applied to music as clusters of *glissando*. The new technical developments of steel-reinforced concrete led to the development of new forms of architecture, such as the ruled surfaces of hyperbolic paraboloids. The form of the hyperbolic paraboloid is another example which was transferred to architecture as well as music in the work of Xenakis. Glissandi in *Metastasis* appear in graphical scores, and we can see the correspondence of the musical form with the architectural form of the Philips Pavilion designed by Xenakis and Le Corbusier for the 1958 world exhibition in Brussels. Later Xenakis founded the "Centre des Etudes Mathématique Automatiques Musicales," where he continued working on the formalisation of aesthetics.

These ideas influenced the student projects "Space-Sound-Sphere", "Balanced Sound Sculpture", "Sphere Music" and "Motion".

"Sound–Sights" projects

The following projects will give an idea of the interdisciplinary working of the student groups and the results presented in the concert and exhibition on 31 October 2003 in Kaiserslautern and later presented in video on a DVD. The concert hall in Kaiserslautern is a neo-Renaissance building built by August von Voit in 1843-46. In one of the projects the students analysed the proportions and history of the building and transformed it in music and visual impressions.

The students were supported and directed by composer Johannes Fritsch of the Music Academy of Cologne, by Norbert Christmann from the Department of Mathematics of the University of Kaiserslautern and Cornelie Leopold from the Department of Architecture of University of Kaiserslautern. The various projects were brought together under the idea of hearing and seeing, music and geometry, combined by signals and short pieces for flute entitled *Pirinore* (Korean for "flute tunes"), according the dramaturgy of the "Sound-Sights" projects. Some of the compositions by Eunshin Jung referred to each other symmetrically in time and space, while others served as bridges for combining several pieces.

Space-Sound-Sphere. The form of music was brought together with the form of an architectural object in this interdisciplinary project of students of architecture and civil engineering together with a student of musical composition. Kim Ngoc Tran Thi composed the pointed sounds for the instrument "Chan". The idea was to contrast the pointed music with a round and smooth spatial form. It had to be possible for a visitor to enter the form and hear the music, so that the experiences of seeing and hearing form are parallel. By contrasting the pointed form of the music to the rounded form of the architecture, the perception becomes more intense. The students decided to build a spherical surface. Several construction and design concepts had been discussed in reference to the costs and possible realisation in a short time as well. Students Philipp Jünemann, Artur Jungiewicz, Jochen Gross and Tobias Wittig developed a concept for building the sphere as a hollowed cube out of Styrodur, sponsored by the chemical company BASF. They built the hollowed cube out of layers, calculated the radius of the circle for each layer and cut the material with by compasses made with a hot wire. The "Space-Sound-Sphere" was built up in the exhibition area of the concert hall so that each visitor was able to experience it individually.

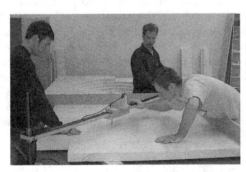

Fig. 6. Building the "Space-Sound-Sphere"

Balanced Sound Sculpture. Another sound installation combined geometric solids with experimental music. Twelve transparent solids were built with speakers inside. Each solid was assigned to corresponding music, bringing the solid into motion. "Music is sound moving form" was the basic idea for this project. The music compositions by Yoon-He Suhmoon, consisting of six motifs for guitar, were assigned to the coloured geometric solids by architecture students Leyla Dal and Filiz Tunc. The moving coloured solids with their shadows produce aural and spatial impressions.

 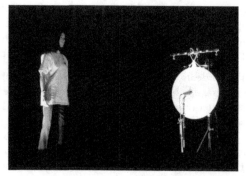

Fig. 7. "Balanced Sound Sculpture" in the Concert Hall and scene from "Sphere Music"

Ball. The most fascinating Euclidean figure is the sphere, which suggests movement through its curved form. The short film project "Ball 2,7" by Dominik Susteck and Joan-Ivonne Bake put a ball in the focus of the film. This object is brought into relation to many other everyday things. The film produced in this way was cut according to certain musical procedures such as repetition, retrograde, inversion, augmentation, etc. The music composer worked with the visual material as he works normally with musical motifs.

Sphere Music. Composer Manfred Ruecker developed his "Sphere Music" from the idea of generating a sphere through a virtual rotation of the area of a tamtam with a diameter of 55 cm. In the rotation process the area was assigned to a volume, and the flat disc became a sphere by rotation. These virtual processes of the sphere formation was converted into musical parameters. Norbert Christmann calculated eleven slices of the sphere with the same volume. These eleven slices of the sphere and eleven relevant pitches of the tamtam formed the basis of the composing process.

Abacus. The "Abacus" picked up the idea of Hermann Hesse's novel *The Glass Bead Game*, in which mathematics and music are joined together by numbers. The name "Abacus" is derived from its outward similarity with the Asian calculators. Architecture student Philipp Jünemann and music composition student Oxana Omeltschuk constructed the "Abacus" with the help of bottles of different sizes. The pitch, timbre and oscillation period of each bottle depends on its form, weight and consistency. The pitch was regulated with aid of the fill height. Coloured water at various fill levels added a visual component to the instrument. An improvisation on the "Abacus" of a composition by music composition student Simon Rummel was performed in the concert hall. There was a second "Abacus" installed in the exhibition area where the visitors were able to try out the instrument on their own.

Point and Line. Points and lines as the basic elements of Euclidean geometry were the musical and visual elements of this project. Built objects, music and graphics expressed points and lines at various sensual layers. Urban and environmental planning student Martin Wisniowski and music composition student Jihyun Kim built their own instruments, five "superstrings" [Gehlhaar 1971]. The superstring is a simple instrument similar to a monochord with two strings over a wooden board and an electromagnetic pickup. Graphical notations of points and lines supported the musical improvisation. The sounds of the superstrings in the concert were visualised in real time by an interactive computer graphics system. The point and line graphics were projected onto two screens to visualise the sounds of the instrument.

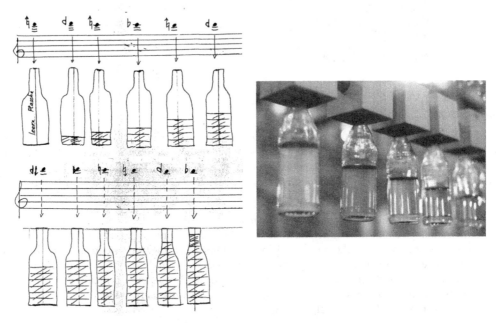

Fig. 8. Realisation and concept of "Abacus"

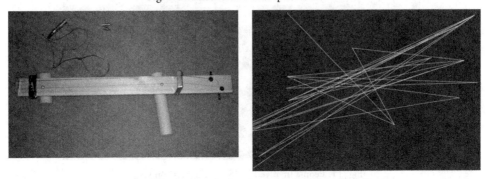

Fig. 9. The "superstring" and graphic "Point and Line"

Motion. Motion, music and geometry were combined in this project in an interactive stage system. Architecture students Nils Hücklekemkes and Pierre Wettels, together with music composition student and dancer Oxana Omeltschuk, developed this stage system, which was able to produce sounds and images in real time. The motion of the dancer on the stage generates music and images through video tracking. The dancer is the composer of music and images at the same time. The interactive stage system allows a dancer to produce sounds solely through his/her actions and at the same time to experience the motion geometrically. With this instrument a total audio-visual artwork was created.

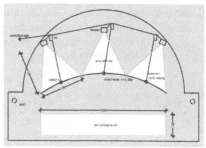

Fig. 10. Concept of "Motion"

Conclusion

It was a challenging task to combine theoretical and practical work and to realise the ideas 1:1 in the concert and the exhibition. The relationships between geometry, architecture and music were studied in theory and practice, realised between science and art, and translated into aural and visual perceptions. Geometry as a structural science is able to connect music and architecture, and music and art.

Working in the interdisciplinary groups made it clear that there are strong reciprocal misunderstandings about other disciplines. It was not easy to bring together the creative ideas of students from the various backgrounds and to mediate between them. Sometimes it was hard to bring the students to the point of being open to other ways of thinking, to understand each other and to combine the different ideas. We did not succeed in all cases in convincing them to work together on the projects. Some projects remained individual works, but were connected with the other projects through the process of the concert.

Fig. 11. People involved in "Sound-Sights" on stage

Finally the concert, the exhibition and the DVD offered the chance to present the projects to a large non-academic audience. Art and science turned out be partners in the mediation process. "Sound-Sights" was realised thanks to the help of many partners, sponsors and technical staff.

References

BALTENSPERGER, André. 1996. *Iannis Xenakis und die stochastische Musik.* Bern: Verlag Paul Haupt.

CHRISTMANN, Norbert. 2003. Töne und Figuren – Über das Zusammenspiel von Geometrie und Musik. Pp 11-26 in *Geometrie, Architektur und Musik,* Cornelie Leopold and Norbert Christmann eds. Technische Universität Kaiserslautern.

GEHLHAAR, R. 1971. Superstring. *Feedback Papers* **2,** Cologne: Feedback Studio Verlag.

HAY, Denys. 1968. *Die Renaissance.* Droemer Knaur München/Zürich. http://music.washcoll.edu/

LEONHARDT, Jochen and Jürgen WILLENBACHER. 2003. Abbildungsgeometrische Analyse musikalischer Formen. Pp 63-76 in *Geometrie, Architektur und Musik,* Cornelie Leopold and Norbert Christmann eds. Technische Universität Kaiserslautern.

LEOPOLD, Cornelie, ed. 2003a. *Klangsichten 311003. Musik sehen - Geometrie hören.* Technische Universität Kaiserslautern, Germany.

———2003b. *Geometrie als Grundlage für Architektur und Musik bei Iannis Xenakis.* Pp. 85-101 in *Geometrie, Architektur und Musik,* Cornelie Leopold and Norbert Christmann eds. Technische Universität Kaiserslautern.

LEOPOLD, Cornelie, Johannes FRITSCH and Andrea EDEL, eds. 2004. *Klangsichten 311003 DVD. Musik sehen - Geometrie hören.* Technische Universität Kaiserslautern. With English subtitles.

VON NAREDI-RAINER, Paul 1982. *Architektur und Harmonie – Zahl, Maß und Proportion in der abendländischen Baukunst.* Cologne: DuMont Verlag.

WITTKOWER, Rudolf. 1971. *Architectural Principles in the Age of Humanism.* W.W. Norton and Company.

About the author

Cornelie Leopold, Academic Director, is a Lecturer and Head of the Descriptive Geometry and Perspective Section of the Department of Architecture, Urban and Environmental Planning and Civil Engineering of the University of Kaiserslautern in Germany. She studied mathematics, philosophy and German at University of Stuttgart. She has published several books and articles on the relationships between geometry and architecture. Her book *Geometrische Grundlagen der Architekturdarstellung* (Kohlhammer Verlag Stuttgart 1999, second edition 2005) introduces the geometry of architectural forms and representations. She is president of the German Society for Geometry and Graphics (Deutsche Gesellschaft für Geometrie und Grafik). Her research interests are descriptive geometry and new media, geometric space conceptions and architecture, visualisation of architecture, development of spatial visualisation skills, and geometry and creation. More about her work can be found at http://www.uni-kl.de/AG-Leopold, and http://www.klangsichten.de.

Book Reviews

Giulio Magli

Misteri e scoperte dell'archeoastronomia
Il potere delle stelle dalla preistoria all'isola di Pasqua

Rome: Newton & Compton Editori, 2005

Reviewed by Manuela Incerti

Università di Ferrera, Dipartimento di Architettura
Via Quartieri,8
44100 Ferrara ITALY
manuela.incerti@unife.it

Questo libro parla di come l'uomo ha vissuto il suo rapporto con il cielo e con le stelle (This book addresses how man has lived his relationship with heaven and the stars) (p.13). With these words the author opens his long and detailed account of the great number of material witnesses that attest to the deep, never-abandoned, millennia-long relationship between man and the Cosmos.

The aim of the book is to gather the results of the most revealing and reliable archaeoastronomical research, highlighting from time to time the most original and innovative contributions of that new discipline to the history of mankind in general, and to the history of science in particular.

The 445 of the book describe and illustrate numerous sites distributed over the globe; it is not possible to list all of the book's themes and cases studies, but these include the megalithic civilizations of Northern Europe, Malta, Egypt, Mesopotamia, India, China, Japan, North America, Mexico, Peru and the Easter Islands.

The text begins with a necessarily concise description of the Palaeolithic period, discussing the studies of Michael Rappenglueck on the cave drawings of Lascaux. With the aid of a computer, Rappenglueck reconstructed the position of the stars over the cave at the time that the caves were believed to have been painted (some 15,000 years ago), and in 1998 he proposed an interesting cosmographic interpretation of the scene that gives validity to the existence of some kind of "astronomic thought" in this long-ago age.

Although it is not possible in any case to be certain about these or other material evidence belonging to the Palaeolithic age, a great many studies give credence to the existence of some astronomical intents in building activity beginning as early as the fifth century B.C.

The most important megalithic sites in Europe as well as the scientific relevance of the discovery of radiocarbon dating in identifying the founding dates and successive developments are described with care and thoroughness.

The sites are classified according to classic typologies:

–   Isolated stones, *Menhirs*;
–   Arches composed of three stones, *Dolmen*;

–   Corridors realized with stone slabs and covered with earth mounds, *Barrows*;
–   Stone circles, *Cromlech*;
–   Circles or ovals of stones surrounded by a moat and earthworks, *Henge.*

The first of all sites examined is Karnak (5000-3000 B.C.). The linear geometries and the extraordinary perspective effects are discussed, as are other material witnesses present in the area (cromlech, dolmen, corridors, great menhir) that confirm the fervent building activity aimed at constructing what the author defines as a "sacred landscape:"

> By sacred landscape we mean an environment in which man lives, made by man himself, observed, built, chosen, created, thought out in accordance with an idea, a mental, religious, scientific or philosophic scheme. However, the specific ways and forms in which this intent is expressed can be completely different from culture to culture...

In the second chapter the author describes the sites of Newgrange, Stonehenge, Avebury and Silbury. Studies undertaken on these places begin with the extraordinary intuitions of Norman Lockyer around the end of the 1800s.

In the course of the second chapter the author examines more "dated" conclusions that have been revised over time, such as those of C. Newham and G. Hawkins, up to the theoretical resolution of the discipline thanks to Alexander Thom, extensively described and commented both in its positive aspects as well as in those that are "weaker" or open to criticism.

The third chapter addresses the well-known theme of the remains or traces of more than forty megalithic temples in Malta. The construction of these buildings, datable to 3,500 B.C., appears to have taken into account a south-eastern orientation; according to Klaus Albrecht (2001). This direction is related to the rising of the sun at the winter solstice. Here as in various other occasions (such as the discussion of the temples of the Easter Islands), the author addresses the practical problem of the transportation and positioning of the huge stones of which the monuments are built, referring from time to time to studies in the history of technology and constructive techniques (the second appendix is dedicated to this).

Particularly fascinating, and perhaps less well known, are the studies on the Monumental Mausoleum of Qin in China. This is a funerary site whose square tomb measures 500 meters to a side, probably the largest ever built. In 1974 three huge trenches were found to the east of the tomb. The largest, 210 meters long and 60 wide, comprises nine parallel corridors about three meters wide each, where 8,000 life-size terracotta statues stand. This is the famous "terracotta army" of the Emperor Qin, who died in 210 B.C. The whole complex, including the rows of warriors, is rigorously oriented towards the cardinal points, and appears to follow a principle inspired by, to paraphrase Magli, a sort of cosmic order between Earth, the plane of man and the heavens, in which Qin placed himself in the centre as a kind of guarantor.

The long list of studies, cited in chronological order and in accordance with their historic evolution, might be somewhat tiring to the neophyte, but testify to the author's commitment to a thoughtful compilation of an extended anthology of studies, a synthesis that was lacking in the panorama of Italian scientific studies.

In the first appendix, "Il cielo ad occhio nudo" (The heavens seen by the naked eye), the reader with less experience in position astronomy will find all the indispensable elements for understanding the phenomena described.

The author-date system of references, normally used in scientific and mathematical work to lighten the load of endnotes or footnotes, will not be familiar to those working in historical, artistic or literary fields.

The tone of the text, sometimes colloquial, transmits the author's participation in the themes that he deals with and his efforts to involve the reader both critically and emotionally.

The book includes indexes of names and sites. The bibliography, rich and up-to-date, is essential for further study.

About the reviewer

Manuela Incerti is Professor or architectural drawing and representation at the Architecture Department of the University of Ferrara, Italy. She got her Ph.D. at the University of Florence, discussing a dissertation that addressed the relationships between astronomy and architecture in the Middle Ages. The research that she actually carries on focus on the theoretical and practical aspects of measuring and surveying monuments, and on the interpretation of the collected data. A member of the Ferrara University team that has been inquiring into 3D scanner survey procedures for some years, she pursues her own personal research on the "design of light" in historical architecture. She is the author of more than forty papers, and of a monograph entitled *Il disegno della luce nell'architettura cistercense. Allineamenti astronomici nelle abbazie di Chiaravalle della Colomba, Fontevivo S. Martino dei Bocci in Valserena.* (Design of Light in Cistercian architecture: Astronomical Orientation of the Abbeys of Chiaravalle della Colomba and Fontevivo S. Martino dei Bocci in Valserena). She teaches courses in measured surveys and drawing, both free-hand and computer-aided, and also drawing for industrial design and fashion.

Edward B. Burger
Michael Starbird

Coincidence, Chaos and All That Math Jazz

New York: WW Norton, 2005

Reviewed by Kim Williams

Nexus Network Journal
Via Cavour, 8
10123 Turin (Torino) Italy
kwilliams@kimwilliamsbooks.com

Coincidence, Chaos and All That Math Jazz presents selected mathematical concepts in an appealing, non-intimidating and amusing manner. Its aim is not mathematical entertainment, but rather to show math at its counter-intuitive best and to vanquish math anxiety. It addresses the ubiquity of mathematics in our everyday world – a subject much talked about, and recently discussed in Turin in an aptly-named and well-attended lecture by Alberto Conte, "The Explosion of Mathematics." Mathematics is present where we expect it and can see it – cryptology, computer technology, in art and architecture – and where we don't expect it and can't see it: in the advanced technology of cell phones, in traffic control, in genetics. Mathematicians are delighted, and justly so, by the attention given to mathematics and mathematicians in the popular media, reflecting the "sex appeal" of mathematics as presented in movies such as "Good Will Hunting" and "A Beautiful Mind" and in books such as *The Da Vinci Code*. *Coincidence, Chaos and All That Math Jazz* is similar in concept to *Math and the Mona Lisa*, reviewed in vol. 7 no. 2 of the Nexus Network Journal, in that it provides a look inside ideas that have been superficially and therefore perhaps misleadingly treated in other popular contexts. Books such as these are relevant for the Nexus reader mainly because they almost invariably discuss the Golden Section, whose accuracies and inaccuracies of application are ever the source of debate. In this case, the discussion of the Golden Mean is an interesting case study.

The discussion starts with asserting the visual beauty of the golden rectangle, then goes into the Fibonacci series, discussed in the previous chapter, then finds a golden rectangle in the Parthenon and in a Grecian eyecup. Alarms started going off in my head already when I saw the figure of the Parthenon, and became louder at the figure of the eyecup, where the key points for the proportions of ϕ had nothing to do whatsoever with how the artisan crafted the object. On the very next page, however, the authors make good, by pointing out the subjectivity of the coincidences of the sides of the determining rectangle with significant elements of the Parthenon (for instance, their rectangle rests on the second step of the base; had it rested on the top step, of course, it wouldn't have fit. The authors might have done well to point out that any number of rectangles could be drawn relative to the Parthenon in a variety of scales, and then refer to Chapter One, their own chapter on randomness, to state that if enough rectangles are generated, some of them will certainly be Golden Rectangles. As Livio Volpi Ghirardini and Marco Frascari pointed out in their presentation at Nexus 1998, this is all the proof that ϕ-lovers need; it's only a problem for those of us who remain sceptical. Having raised a red flag as to the subjectivity of

1590-5896/06/010138-2 DOI 10.1007/s00004-006-0009-7

finding the Golden Section, the authors then leave it up to the reader to decide whether or not there are Golden Rectangles in paintings by Seurat, Mondrian and Leonardo. I thought they somehow shirked their responsibility here.

The presence or not of the Golden Mean in music is also discussed, but pretty unconvincingly: "The length of the entire prelude is 129 seconds, and the length of time between the beginning of the piece and that fantastic fortissimo at bar 70 is 81 seconds. Dividing those two quantities yields 1.592..., which is impressively close to the Golden Ratio of 1.618..." (p. 130). But is it indeed impressively close? It is a 1.6% error; in such a small measure as 129, 80 would have been a closer approximation than 81. The 1.592 value argues more strongly for the ratio of 5:8 than it does for the Golden Section. We have heard this kind of argument many a time in the analysis of architectural monuments; it simply points to the relative impossibility of affirming the intentional use of the Golden Mean as a design or compositional tool, unless, of course, the architect or composer is living and can state his intentions.

It's those *formulae* that are so off-putting in mathematics, isn't it? If we could just get rid of the formulae, math would be ever so much more appealing. I discussed this in my review of *Math and the Mona Lisa*, where the formulae were relegated to the appendix in order not to bore or intimidate the lay reader. Here authors Burger and Starbird have tried very hard – and have almost entirely succeeded – to eliminate formulae. This leads to some paragraphs of this type: "First we note that 3 squared is 9, minus 2 yields 7. Next 7 squared is 49, minus 2 is 47." Is that really easier and less intimidating than "$3^2-2=7$, and $7^2-2=47$"?

In my mind both the treatment of the Golden Section and the attempt to eradicate formulae raise the question of rigor. In order to make the mathematical concepts clear, rigor is often sacrificed (if there is anything besides a formula to turn off the layman's interest, it is a proof). But to mathematicians, it is precisely the rigor that leads to beauty in mathematics. It could perhaps be explained to the layman as similar to listening to Ella Fitzgerald sing scat: as long as she starts out slow, most of us can hum or sing along, but when she really gets going, very very few can keep up. But that doesn't mean that Ella's music isn't incredibly beautiful: you wouldn't want her to limit herself to singing just the easy parts so that we mortals can do it too.

One last disappointment is there are no indications as to further reading on the topics presented. A bibliography would have been very useful.

In spite of my reservations, however, there is plenty of interesting material here, and it is entertainingly presented. What comes across very clearly is that mathematics is not as mysterious and forbidding as it seems: it is rather based on very clear thinking. Those who strive valiantly to make math interesting to indifferent students will find inspiration in this book; it would be very well recommended to students who ask, "Why would I want to learn mathematics?" Many of the topics presented are relevant to Nexus themes: the Fibonacci series, continued fractions, tiling of the plane, fractals, topology, the fourth dimension.... themes that we always try to approach with scientific rigor, and/or searching for the right balance between rigor and popularization when didactics is concerned.

About the reviewer

Kim Williams is editor-in-chief of the *Nexus Network Journal*.